DEVELOPMENT OF WASTE ACCEPTANCE CRITERIA FOR LOW AND INTERMEDIATE LEVEL RADIOACTIVE WASTE

The following States are Members of the International Atomic Energy Agency:

AFGHANISTAN
ALBANIA
ALGERIA
ANGOLA
ANTIGUA AND BARBUDA
ARGENTINA
ARMENIA
AUSTRALIA
AUSTRIA
AZERBAIJAN
BAHAMAS, THE
BAHRAIN
BANGLADESH
BARBADOS
BELARUS
BELGIUM
BELIZE
BENIN
BOLIVIA, PLURINATIONAL
  STATE OF
BOSNIA AND HERZEGOVINA
BOTSWANA
BRAZIL
BRUNEI DARUSSALAM
BULGARIA
BURKINA FASO
BURUNDI
CABO VERDE
CAMBODIA
CAMEROON
CANADA
CENTRAL AFRICAN
  REPUBLIC
CHAD
CHILE
CHINA
COLOMBIA
COMOROS
CONGO
COOK ISLANDS
COSTA RICA
CÔTE D'IVOIRE
CROATIA
CUBA
CYPRUS
CZECH REPUBLIC
DEMOCRATIC REPUBLIC
  OF THE CONGO
DENMARK
DJIBOUTI
DOMINICA
DOMINICAN REPUBLIC
ECUADOR
EGYPT
EL SALVADOR
ERITREA
ESTONIA
ESWATINI
ETHIOPIA
FIJI
FINLAND
FRANCE
GABON
GAMBIA, THE

GEORGIA
GERMANY
GHANA
GREECE
GRENADA
GUATEMALA
GUINEA
GUYANA
HAITI
HOLY SEE
HONDURAS
HUNGARY
ICELAND
INDIA
INDONESIA
IRAN, ISLAMIC REPUBLIC OF
IRAQ
IRELAND
ISRAEL
ITALY
JAMAICA
JAPAN
JORDAN
KAZAKHSTAN
KENYA
KOREA, REPUBLIC OF
KUWAIT
KYRGYZSTAN
LAO PEOPLE'S DEMOCRATIC
  REPUBLIC
LATVIA
LEBANON
LESOTHO
LIBERIA
LIBYA
LIECHTENSTEIN
LITHUANIA
LUXEMBOURG
MADAGASCAR
MALAWI
MALAYSIA
MALI
MALTA
MARSHALL ISLANDS
MAURITANIA
MAURITIUS
MEXICO
MONACO
MONGOLIA
MONTENEGRO
MOROCCO
MOZAMBIQUE
MYANMAR
NAMIBIA
NEPAL
NETHERLANDS,
  KINGDOM OF THE
NEW ZEALAND
NICARAGUA
NIGER
NIGERIA
NORTH MACEDONIA
NORWAY
OMAN

PAKISTAN
PALAU
PANAMA
PAPUA NEW GUINEA
PARAGUAY
PERU
PHILIPPINES
POLAND
PORTUGAL
QATAR
REPUBLIC OF MOLDOVA
ROMANIA
RUSSIAN FEDERATION
RWANDA
SAINT KITTS AND NEVIS
SAINT LUCIA
SAINT VINCENT AND
  THE GRENADINES
SAMOA
SAN MARINO
SAUDI ARABIA
SENEGAL
SERBIA
SEYCHELLES
SIERRA LEONE
SINGAPORE
SLOVAKIA
SLOVENIA
SOMALIA
SOUTH AFRICA
SPAIN
SRI LANKA
SUDAN
SWEDEN
SWITZERLAND
SYRIAN ARAB REPUBLIC
TAJIKISTAN
THAILAND
TOGO
TONGA
TRINIDAD AND TOBAGO
TUNISIA
TÜRKİYE
TURKMENISTAN
UGANDA
UKRAINE
UNITED ARAB EMIRATES
UNITED KINGDOM OF
  GREAT BRITAIN AND
  NORTHERN IRELAND
UNITED REPUBLIC OF TANZANIA
UNITED STATES OF AMERICA
URUGUAY
UZBEKISTAN
VANUATU
VENEZUELA, BOLIVARIAN
  REPUBLIC OF
VIET NAM
YEMEN
ZAMBIA
ZIMBABWE

The Agency's Statute was approved on 23 October 1956 by the Conference on the Statute of the IAEA held at United Nations Headquarters, New York; it entered into force on 29 July 1957. The Headquarters of the Agency are situated in Vienna. Its principal objective is "to accelerate and enlarge the contribution of atomic energy to peace, health and prosperity throughout the world".

IAEA NUCLEAR ENERGY SERIES No. NW-T-1.29

DEVELOPMENT OF WASTE ACCEPTANCE CRITERIA FOR LOW AND INTERMEDIATE LEVEL RADIOACTIVE WASTE

INTERNATIONAL ATOMIC ENERGY AGENCY

VIENNA, 2025

© IAEA, 2025

Printed by the IAEA in Austria
December 2025
STI/PUB/2117
https://doi.org/10.61092/iaea.oag0-7yxo

IAEA Library Cataloguing in Publication Data

Names: International Atomic Energy Agency.
Title: Development of waste acceptance criteria for low and intermediate level radioactive waste / International Atomic Energy Agency.
Description: Vienna : International Atomic Energy Agency, 2025. | Series: IAEA nuclear energy series, ISSN 1995-7807 ; no. NW-T-1.29 | Includes bibliographical references.
Identifiers: IAEAL 25-01775 | ISBN 978-92-0-114925-1 (paperback : alk. paper) | ISBN 978-92-0-115025-7 (pdf) | ISBN 978-92-0-115125-4 (epub)
Subjects: LCSH: Radioactive waste disposal. | Radioactive wastes — Management. | Radioactive wastes — Storage. | Low level radioactive waste disposal facilities.
Classification: UDC 621.039.7 | STI/PUB/2117

FOREWORD

The IAEA's statutory role is to "seek to accelerate and enlarge the contribution of atomic energy to peace, health and prosperity throughout the world". Among other functions, the IAEA is authorized to "foster the exchange of scientific and technical information on peaceful uses of atomic energy". One way this is achieved is through a range of technical publications including the IAEA Nuclear Energy Series.

The IAEA Nuclear Energy Series comprises publications designed to further the use of nuclear technologies in support of sustainable development, to advance nuclear science and technology, catalyse innovation and build capacity to support the existing and expanded use of nuclear power and nuclear science applications. The publications include information covering all policy, technological and management aspects of the definition and implementation of activities involving the peaceful use of nuclear technology. While the guidance provided in IAEA Nuclear Energy Series publications does not constitute Member States' consensus, it has undergone internal peer review and been made available to Member States for comment prior to publication.

The IAEA safety standards establish fundamental principles, requirements and recommendations to ensure nuclear safety and serve as a global reference for protecting people and the environment from harmful effects of ionizing radiation.

When IAEA Nuclear Energy Series publications address safety, it is ensured that the IAEA safety standards are referred to as the current boundary conditions for the application of nuclear technology.

To ensure safety, radioactive waste management has to be carried out in a regulated manner compatible with nationally and internationally agreed principles and standards. To achieve this, organizational and administrative arrangements clearly defining the competencies and responsibilities of the responsible institutions involved need to be in place. Radioactive waste management covers all administrative and operational activities that are involved in the generation, pretreatment, treatment, conditioning, transportation, storage and disposal of the various types of waste. The management of waste from 'cradle to grave' is therefore a stepwise process involving a variety of activities, facilities and organizational arrangements.

Often, the various steps are performed by different organizations; as the waste passes from one organization to the next, so too does the responsibility for safety and the associated liability and risk. Such multiparty arrangements demand careful attention to interfaces and, in the case of radioactive waste management, to the development and documentation of waste acceptance criteria (WAC) as the principal means of ensuring clear interfaces. WAC aim to ensure that activities performed in one waste management step (e.g. waste conditioning) do not preclude any anticipated and necessary activities in subsequent waste management steps (e.g. waste storage or disposal). Waste management facilities such as those for processing, storage and disposal operate under distinct and strictly controlled licence conditions. When waste is transferred from one facility to another, it is essential that all licence conditions of the receiving facility are met.

As established in the IAEA safety standards, WAC are one of the main outcomes of developing a safety case and supporting safety assessments, which demonstrate the operational and long term safety of a waste management facility. In accordance with this safety evaluation, the operator of a waste management facility establishes WAC to inform the waste generator about the types and characteristics of waste that can be accepted. Waste acceptance criteria specify the radiological, mechanical, physical, chemical and biological characteristics of waste packages and unpackaged waste that can be processed, stored or disposed of. Once the WAC are established, other operational factors may lead to the adoption of additional criteria, which may be more restrictive.

The principal entities in radioactive waste management are the waste generator, the predisposal operator and the disposal operator, together with the regulatory body; the interfaces between these entities are supported by the use of WAC.

The aim of this publication is to assist all those involved, directly or indirectly, in radioactive waste management by providing useful, practical information for the development, implementation and application of WAC that specify the waste management organizations' requirements for pretreatment,

treatment, conditioning, transportation, storage and disposal of radioactive waste. Included is information on selecting the most appropriate parameters for inclusion in a waste acceptance criterion, as well as on setting and justifying the particular limits or acceptable range of values for a given parameter.

The IAEA is grateful to Z. Drace (Canada), M.I. Ojovan (United Kingdom), P. Ormai (Hungary), A. Morales Leon (Spain) and all others who took part in the planning and preparation of this publication. The IAEA officers responsible for this publication were F.N. Dragolici and G.H. Nieder-Westermann of the Division of Nuclear Fuel Cycle and Waste Technology.

CONTENTS

1. INTRODUCTION

1.1. BACKGROUND

Radioactive waste management (RWM) is a necessary consequence of the application of nuclear energy and technology for research, power generation, the production and use of radioactive sources, nuclear fuel production and reprocessing, decommissioning of nuclear facilities and the remediation of areas contaminated resulted from past nuclear activities. It is a stepwise process that covers all administrative and operational activities involved in the predisposal and disposal of radioactive waste.

The IAEA Safety Standard Series provides a set of principles, requirements and guidance, which are to be considered and applied to RWM programmes within IAEA Member States. As established in the IAEA Fundamental Safety Principles [1], **"The fundamental safety objective is to protect people and the environment from harmful effects of ionizing radiation."** This objective applies to all circumstances from which radiation risks can arise and is relevant to all facilities and activities involving nuclear materials and for all stages over the lifetime of a facility or source of ionizing radiation. Several underpinning safety standards [2–13] specify the requirements and responsibilities of the various parties involved in nuclear activities, including waste generators, waste processors and waste storage and disposal implementers. These organizations are typically involved in the development and application of waste acceptance criteria (WAC), which are an essential aspect of RWM. The IAEA safety standards provide guidance in the interpretation and implementation of these requirements and responsibilities. The IAEA safety standards [8, 9] also have requirements for the development of WAC and its position in overall RWM. These standards position WAC based on the safety case for the facility or are included in the safety case as part of the operational limits and conditions and controls.

The Joint Convention on the Safety of Spent Fuel Management and on the Safety of Radioactive Waste Management is a legally binding treaty that demonstrates a commitment by the contracting parties to achieve and maintain a consistently high level of safety in the management of spent fuel and radioactive waste [14]. Where relevant, information on WAC development and implementation is an important component of the national reports submitted by Member States as part of the Joint Convention.

WAC are most commonly defined as "quantitative or qualitative criteria specified by the *regulatory body*, or specified by an *operator* and approved by the *regulatory body*, for the *waste form* and *waste package* to be accepted by the *operator* of a *waste management facility*" [15]. They specify conditions that are to be met and may include the exclusion of certain materials or limits placed on the content of materials within waste packages and for a facility as a whole. The derivation of appropriate limits and specifications for acceptable waste packages are an important consideration based on facility design, safety assessment and safety case development. Requirements specified by WAC, as well as the methodologies employed to derive and apply these criteria, are documented and approved in line with international standards of safety, best practices and recommendations.

Generally, WAC are either prescribed by the relevant national regulatory authorities or derived from safety assessments performed by facility operators in line with appropriate guidelines defined in the safety standards [10]. Limits and conditions of particular importance for a facility or activity are the acceptable waste inventory and/or the concentration levels for specific radionuclides in the waste. These need to be defined on the basis of the results of the safety assessment supporting development of the safety case. In addition to radiological properties, WAC need to be defined in terms of mechanical, physical, chemical and biological properties or any other applicable characteristics for waste packages or unpackaged waste. Waste acceptance criteria for the facility may be established both for individual waste packages and for the facility as a whole. At a minimum, radionuclide content, radiation safety parameters and quality parameters need to be considered when deriving such requirements.

It is worth noting that WAC development is influenced by every step in the waste management process. Furthermore, WAC play a key role in the management of the interfaces between facility operators involved in waste generation, waste processing, storage and, importantly, disposal. This arises from the use of WAC as a mechanism to assess waste acceptability in relation to the specific conditions pertaining to a given waste management step and the licensing conditions of an involved facility.

This emphasizes the need for a careful and methodical development of well defined, clear, practicable and verifiable WAC, developed through dialogue and consultation with all relevant stakeholders, and with an understanding of the implications the specifications in a WAC may have on waste processing and waste form development to meet the criteria. It is essential that waste processing and waste storage facilities develop their acceptance criteria as far as practicable, taking into account the requirement for waste disposability.

Complementing WAC development is the acceptance process that highlights the importance of the waste characterization and conditioning process qualification, together with checks and controls on waste receipt to verify compliance.

1.2. OBJECTIVE

The objective of this publication is to provide an approach for the development and implementation of WAC for the steps to be undertaken in the planned RWM cycle. By selecting and addressing the relevant operational and safety parameters, this publication will constitute a valuable resource for all those involved in the development and implementation of appropriate WAC, such as waste generators, processors and waste package producers, as well as regulators, RWM organizations and technical support organizations.

Guidance and recommendations provided here in relation to identified good practices represent expert opinion but are not made on the basis of a consensus of all Member States.

1.3. SCOPE

This publication covers the development and implementation of WAC for low and intermediate level waste during all stages of RWM. It describes why WAC are important and which entities are responsible for their development and use at the various stages of waste management. The approach described is largely independent of the type of facility (processing, storage or disposal) because this dependence is supposed to be addressed in the safety assessment and the development of the safety case for the facility considered. The approach is valid for very low level waste (VLLW), very short lived waste (VSLW), low level waste (LLW) and intermediate activity level short lived waste including unpackaged waste (e.g. parts of decommissioned steam generators or pressure vessels) and can be applied to the management and disposal of disused sealed radioactive sources (DSRS) declared as radioactive waste.

The approach for developing WAC for high level waste (HLW) and spent nuclear fuel declared waste or for mining and milling waste is not within the scope of this publication.

1.4. STRUCTURE

This publication provides an approach to establishing a stepwise methodology for the development and revision of WAC, from qualitative to quantitative criteria in accordance with the needs established by the waste inventory to be managed; it ends with WAC for disposal.

Sections 2 to 6 (as well as the appendices) of this publication present technical content. Examples of WAC for disposal taken from various national programmes (and one IAEA programme) are provided in a series of annexes.

The technical content of Section 2 addresses the role of WAC in RWM; it describes the relationship of WAC with the overall waste management system; and introduces the importance of the waste inventory and characterization of the waste streams during the development of WAC. Section 2 also summarizes and compares the role of WAC for different waste management facilities, from generation to disposal.

Section 3 covers stakeholder responsibilities with respect to WAC. It identifies particular responsibilities for waste generators, waste processors, waste storage and disposal operators, regulators and the national government. It also points out common responsibilities for all stakeholders and provides a summary of responsibilities for the preparation, communication and retention of waste records and for maintaining a suitable quality management system (QMS).

Section 4 describes the approach for the development of WAC and details the content and structure of the WAC document.

In Section 5, the implementation of the waste package acceptance process is described, covering the establishment of an acceptance agreement, qualification of waste characterization, production and quality assurance processes, controls and the handling of departures and non-conformities. Section 6 contains the conclusions of this publication.

Appendices I–V provide more information as follows:

— I. Fulfilling waste package acceptance criteria;
— II. Handling departures and non-conformities;
— III. Examples of waste acceptance requirements for predisposal;
— IV. Waste containers that meet acceptance criteria;
— V. An example of a WAC document.

Annexes I–VIII provide examples of WAC for disposal based on experiences in Belgium, France, Japan, Slovakia, Spain, Hungary, Pakistan and for the IAEA system for borehole disposal of DSRS.

2. ROLE OF WASTE ACCEPTANCE CRITERIA IN WASTE MANAGEMENT

2.1. WASTE MANAGEMENT SYSTEM

It is essential that any organization involved in the operation of a RWM facility or its regulation undertake its responsibilities within a comprehensive, integrated and robust management framework (i.e. a management system) comprising a set of interrelated elements for establishing policies and objectives and for enabling objectives to be achieved in a safe, efficient, and effective manner. According to IAEA safety standards, a management system needs to be established, implemented, assessed, continually improved and applied to all stages of RWM [16].

The role of a RWM system is to ensure (1) the protection of the public and the environment against the potential hazards arising from radioactive waste [16]; and (2) the implementation of the system under an appropriate regulatory process according to international standards [6]. The primary objective of this system is to ensure safety, which requires organizational and administrative arrangements to define and implement the activities to be performed. The system includes ensuring the availability of sufficient capability and capacity for carrying out the activities [6]. The assurance of safety is to be justified in a safety case that is achieved primarily through the application of two safety functions: the containment of radioactive materials to prevent or control their release and dispersion, and through isolation, a function of engineered and natural barriers that provide for the separation of radioactive waste from humans and the wider environment [10].

The basic steps in RWM are handling (collection, characterization and segregation), then pretreatment, treatment, conditioning (collectively termed processing), storage and disposal. The interdependencies between steps are such that decisions on RWM made at one step may preclude or otherwise affect alternatives at a subsequent step. Conflicting requirements that could compromise

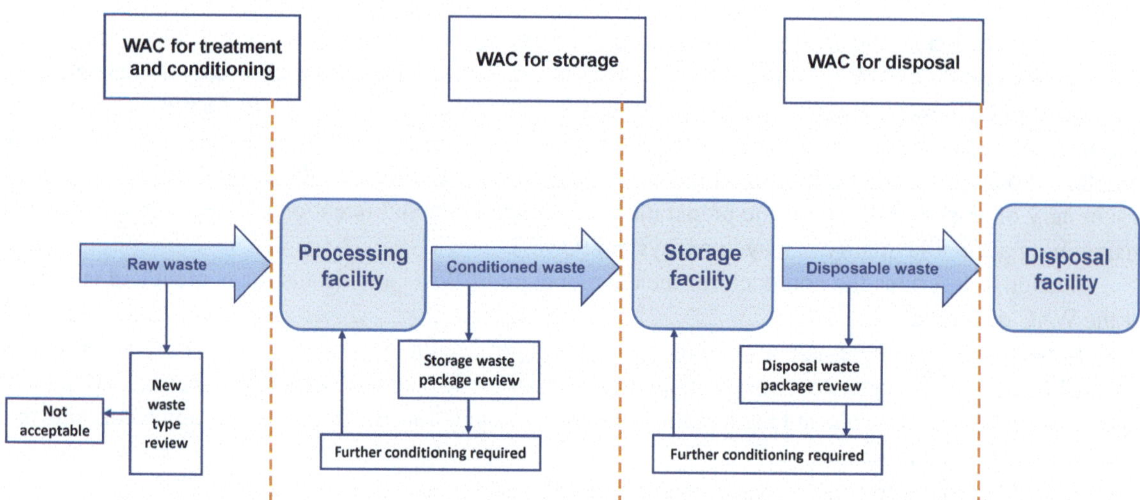

FIG. 1. The interfaces between different waste management facilities.

operational and long term safety at any stage need to be avoided [1]. The various steps will have to be well coordinated (with responsibilities clearly defined in the takeover agreements). The interfaces between steps are important because as the waste passes from one facility to another, so too does the responsibility for safety and the associated liability and risk.

WAC are used as a key tool for managing the overall process of waste transfer between entities and are an essential pillar of a waste management system, together with qualified staff and approved procedures, effective processes, defined and well characterized waste packages, and well run, licensed facilities. WAC may be established for a specific waste form, individual waste packages and/or for a facility as a whole.

The accepted inventory levels relating to volume, waste form and the activity associated with specific radionuclides within a waste package or facility are, in most cases, reliant on the assessment of various scenarios and on established criteria associated with predisposal waste management activities and waste minimization activities such as discharge and clearance [6]. Figure 1 illustrates the interfaces between the various waste management facilities and the role of WAC in the movement of waste between facilities and/or organizations. Section 2.2 provides further details of how WAC are used in the interface management of waste movements between different waste management facilities.

The interfaces in Fig. 1 are simplified for the purposes of illustration. In some Member States, there may be a degree of vertical integration. Assumptions here are that the waste generator is responsible for waste collection and segregation; the operator of the processing facility is responsible for receipt of the waste, its processing and subsequent temporary retention until sending for further management steps, the operator of storage facility is responsible for receiving waste packages, their storage, subsequent retrieval and transfer to a disposal facility; and the operator of disposal facility is responsible for receiving waste packages and placing the packages in a repository. The responsibility for the transport of waste may reside with the facility operator (either the organization involved in dispatching or receiving the waste) or with a third party. Consequently, while facility-specific WAC typically include requirements for the transport of waste, in some circumstances dedicated WAC for transportation alone may be necessary.

Some examples where WAC are relevant in the RWM process include:

— LLW arising from nuclear power plant operation and maintenance requiring conditioning, storage and eventual disposal: WAC need to be specified by the conditioning, storage and disposal facilities and earlier management stages need to consider WAC for all subsequent steps.

— LLW from fuel reprocessing (effluent stream collection, conditioning to an immobilized waste form and storage): The role of WAC is to ensure that the waste form is suitable for the storage configuration, and WAC for eventual disposal need to be anticipated as much as possible.
— DSRS (recovery/collection, interim storage, conditioning, possible further storage and disposal): Generic criteria might be appropriate for providing safety during the storage of unconditioned DSRS, but subsequent processing, further storage and disposal require the development of WAC for processing, WAC for storage and WAC for disposal to provide safety at each of these stages (see Annex VIII).

Because the implementation of RWM is typically associated with a long period between waste generation and disposal, there are, in practice, many situations where decisions need to be made before all radioactive waste management activities are established. As far as reasonably practicable, the effects of future radioactive waste management activities, particularly disposal, need to be anticipated and taken into account when any preceding radioactive waste management activity is being considered [8, 9].

2.2. ROLE OF WASTE ACCEPTANCE CRITERIA

Adequate RWM is highly dependent on the quality of the design and construction of the infrastructure (predisposal or disposal) to meet the associated requirements related to the context of the safety case, as well as on the properties and characteristics of the waste. The safety case represents an integrated consideration of safety for each of the RWM steps, addresses the operational safety of the facility and the interdependences and interconnections among all radioactive waste management steps. The acceptability of the waste forms will be judged based on compliance with the WAC established for all subsequent RWM steps, in particular processing, storage, transport and if available, disposal of the waste [5, 8]. In order for the waste to be safely managed, both in the short and the long term, WAC have to comply with specific and appropriate requirements to ensure acceptance into the respective facility, whether for treatment/conditioning, storage or disposal. If the waste accepted by a facility fully meets properly set WAC that conform to licensing conditions and the safety case, then there is assurance that known risks to the facility arising from the waste packages have been adequately assessed. WAC are therefore essential to the waste management system, together with qualified or approved procedures, processes employed, defined packages and the safety cases of the licensed facilities involved. WAC provide guidance for determination of upstream processes and for controls at waste acceptance. Therefore, the existence of WAC can be considered as comprising a key part of the overall waste management process, providing the basis for the interface agreements between dispatching and receiving waste management facilities. It is advisable to establish WAC for each stage of the various activities in a waste management system. Operators will develop their WAC based on their permitted waste inventory, waste management strategy and the selected end point option. The developed WAC will become operational only once they are approved by the regulatory body.

WAC are important in the RWM for several reasons:

— WAC ensure compliance with relevant safety requirements throughout RWM.
— WAC can prevent or minimize technological problems that might otherwise arise later during RWM.
— WAC assist the waste generators and facility operators to effectively discharge their obligations in managing radioactive waste and in selecting appropriate technologies and processes.
— WAC assist standardization of waste management operations.
— WAC are essential to ensure adequate waste tracking.

Besides their main purpose, WAC are also a vital point of reference for communication and dialogue between all stakeholders involved in RWM. They ensure the safe and efficient transfer, processing, storage and disposal of wastes according to clear guidelines provided to upstream management organizations and in accordance with facility design, licensing and safety case requirements.

2.2.1. Relationship between waste acceptance criteria for disposal and predisposal

According to national demands or the national programme, the development of WAC and the development of disposal routes need to be carried out in parallel [8–10, 12]. WAC will be derived from the consideration of both local operational requirements (safety and technical) as well as requirements imposed by site specific safety assessments for a storage and/or disposal facility. It is important to note that these criteria have to be qualitatively or quantitatively based, to facilitate the conformance assessment by direct measurement, calculations and/or by application of appropriate management methods and inspections.

If neither a national disposal facility nor concept are available, consequently adequate waste acceptance requirements for disposal may not exist. Nevertheless, for the proper and safe management of already existing and currently arising radioactive waste, the development and implementation of WAC are necessary for waste processing and storage. For transportation, the requirements are established in Ref. [3]. Predisposal WAC address operational safety requirements and ensure proper implementation of technological processes, and they will comply with the storage/disposal requirements to avoid compromising later management steps.

Acceptance (control) measures will also be applied to non-radioactive materials entering processing technologies to ensure the proper quality of generated waste forms or packages, e.g. type and quality of binding and construction materials and additives. These criteria are usually set in prescriptions for technological procedures.

Where site specific disposal facility information is not yet available, and therefore a full range of disposal acceptance criteria cannot be established, a preconditioning step may be required for acceptance for storage. Such conditioning will be undertaken in a way that ensures extensive rework will not be required when requirements for disposal become available. For instance, straightforward measures such as waste sorting and segregation and treatments such as decontamination or compaction with overpacking of the pucks, but without cement in-filling, would allow for effective storage. In such an instance, however, further conditioning might be required for disposal. Therefore, as far as feasible, such treatments need to keep in focus the range of viable disposal options and ensure, as far as practicable, that an optimum management solution is achieved.

The development of WAC when a disposal route is not defined is discussed further in Section 4.4.1.

2.2.2. The role of waste acceptance criteria throughout radioactive waste management

WAC are either generic or specific to a particular facility or site and may embrace a variety of radioactive waste forms and/or waste packages. The primary role of WAC in relation to typical waste management facilities is summarized in Table 1.

Application of WAC facilitates the following functions in managing the waste transfer process between facilities, with disposal being the eventual goal:

(a) Controlling the appropriate segregation of waste streams into waste types and waste categories that facilitate effective management in downstream facilities;
(b) Defining characterization criteria such that information requirements associated with waste packages can be provided for the purposes of efficient and safe further treatment/conditioning, storage and/or disposal;
(c) Defining waste types and packaging (as well as package performances to be met) such that the packages can be transported, handled, stacked, inspected and retrieved using available equipment and processes;

TABLE 1. SUMMARY OF THE PRIMARY ROLE OF WAC IN RELATION TO THE TYPICAL WASTE MANAGEMENT FACILITIES

Activity or facility	Description	Applicable waste types	Primary role of WAC for facility
Waste generating facility	Minimizes waste generation. Avoids prohibited items in raw waste and follows simple rules to collect raw waste in separate streams depending on the nature of waste generated. Provides initial characterization (chemical, physical and radiochemical) of waste streams.	Raw waste	WAC for processing are considered for preparing the waste for the next management step.
Waste processing	Accepts raw waste and converts it into a form suitable for further management (e.g. incineration, compaction, melting, cementation, packaging). Various types of facilities are utilized depending on the nature of the waste and processing technologies, which usually include: — Pretreatment facilities such as those for decontamination and size reduction; — Treatment facilities used to reduce the volume, remove the radionuclides and change composition including compactors, super-compactors and incinerators for solid waste and water purification facilities for aqueous waste; — Immobilization facilities converting raw waste into stable durable waste forms using technologies such as cementation, bituminization and vitrification; — Containerization and overpacking facilities producing waste packages suitable for safe storage, transportation and disposal.	Raw, pretreated, treated and conditioned waste.	Control and limit the incoming wastes to types that can be handled by the installed equipment and processes, within the prescribed safety limits.
Waste storage	Accepts treated or conditioned wastes for storage either above ground (e.g. storage building) or in-ground (e.g. cavern).	Treated, packaged and/or conditioned waste, according to the type of waste and facility (generally solid or solidified wastes only).	Control and limit incoming waste packages to those that can be safely stored and monitored for the required period of time (protection of workers, the public and the environment). Provide for handling, stacking, inspection and retrieval of waste packages.
Waste transportation	Transport waste between facilities while providing protection for people, property and the environment from the effects of radiation.	Raw, treated and/or conditioned wastes.	Control and ensure containment of the radioactive contents, limit external radiation levels and prevent criticality.

TABLE 1. SUMMARY OF THE PRIMARY ROLE OF WAC IN RELATION TO THE TYPICAL
WASTE MANAGEMENT FACILITIES (cont.)

Activity or facility	Description	Applicable waste types	Primary role of WAC for facility
Waste disposal	Accept waste for disposal. Various types of facilities are possible, depending on the level and nature of the hazard associated with the wastes and a preferred disposal concept: — Authorized landfill disposal: Disposal in a facility similar to a conventional landfill facility for industrial refuse but which may incorporate measures to cover the waste. Such a facility may be designated as a disposal facility for VLLW with low concentrations or quantities of radioactive content. — Near surface disposal: Disposal in a facility consisting of engineered trenches or vaults constructed on the ground surface or up to a few tens of metres below ground level. Such a facility may be designated as a disposal facility for LLW. — Disposal of ILW: Disposal can be performed in different types of facilities located from a few tens of metres up to hundreds of metres below ground (e.g. caverns, silos). — Geological disposal: (a) A mined facility constructed in a particular geological formation (tunnels, vaults or silos) and mandatory at least a few hundred metres below ground level or (b) borehole disposal: can be a single or several boreholes varying from a few tens of metres up to a few hundreds of metres deep (this disposal facility is suitable only for small waste volumes, in particular DSRSs.	Solid, conditioned waste packages, according to the type of facility. Degree of packaging and conditioning will depend on the waste characteristics and the role assigned to packaging and conditioning in the overall design and safety case of the facility.	Control and limit incoming waste packages to those that can meet the operational and post-closure safety requirements of the facility (protection of workers, the public and the environment). Provide for handling and stacking of waste packages.

(d) Controlling and ensuring containment of radioactive contents, controlling and limiting external radiation levels, preventing criticality and damage caused by radiolysis, internal heat generation and/or ambient or adverse environmental conditions;

(e) Ensuring that facility design and performance requirements are met;

(f) Providing for the protection of workers, the public and the environment by ensuring that short term and long term safety requirements are met.

The WAC are specific to a waste stream, waste processing technology and waste processing facility, and typically include a range of requirements with which a facility operator will have to comply. If a waste package meets the appropriate set of WAC, then, by implication, it is certified that it meets all of the design safety and performance requirements of the receiving facility. On that basis, WAC form the basis of an agreement between waste shippers and waste receivers for the satisfactory transfer of waste.

2.2.3. Role of waste inventory and waste characterization

Knowledge of the waste inventory is a prerequisite for the determination of a waste management strategy and for the development of the required WAC. Therefore, information about the inventory will be compiled at the beginning of the waste management process, to the greatest extent possible and as a priority.

The safety assessment of a disposal facility, but also that of any other waste management facility, is driven by the waste inventory. It is usual for knowledge of the inventory to evolve with time. For instance, additional waste streams may be added to the inventory, short lived radionuclides may decay during storage or certain radionuclides may be difficult to measure with accuracy using existing methods. Regular updating of the inventory, with the aim of making it sufficiently robust, will therefore be part of the waste management process.

A waste characterization strategy and corresponding implementation programme represent a first step in the RWM chain. The waste characterization strategy needs to take into account waste characteristics and anticipated WAC that would include consideration of process controls, quality assurance requirements, transportation requirements and worker safety requirements. These aspects are necessary to ensure that planned characterization methods will provide sufficient information to confirm waste compliance. The development of WAC is an iterative process to be developed through consultation, as is discussed in more detail in Section 3. A waste characterization programme aims at providing sufficient data of appropriate quality concerning the waste produced by a waste generator, verifying that process parameters are within the limits of the relevant WAC and providing confidence in all the waste management steps based on quality control results. This programme will consider the historical, current and future plans for the waste including:

— Originating process (nuclear power plant, research reactor, fuel reprocessing, historical waste, etc.);
— Conditioning process together with the waste package components (e.g. immobilization matrix, drums, metal boxes, etc.);
— Location(s) where the wastes are stored or disposed;
— The recovery, collection and processing of existing data;
— Waste characteristics.

The characterization of waste packages requires information concerning:

— Radiological content (key nuclides, activity, scaling factors for difficult-to-measure nuclides);
— Chemical properties (composition, concentration, etc., and including toxic species);
— Physical/mechanical properties of the waste form and/or waste package (form, compressive strength, homogeneity, etc.);
— Thermal properties;
— Radionuclide mobilization characteristics (gaseous and water paths, i.e. gas generation and leachability);
— Biological properties.

The waste characterization programme will incorporate a sampling and measurement/analysis plan including not only information concerning the techniques and equipment for sampling, measurements and analysis, but also the methods to be used to manage the recording of all such information, as well as the uncertainties associated with key parameters. The strategy and methodology to be followed in establishing and implementing a characterization programme for the various kinds of radioactive waste streams or packages are discussed in detail in Ref. [17], which addresses two important themes: the first is dealing with the proper strategy related to waste characterization and control, while the second is providing a description and evaluation of selected control procedures and methods for all basic steps of the radioactive waste life cycle.

3. RESPONSIBILITIES IN RELATION TO WASTE ACCEPTANCE CRITERIA

Safe RWM has to be carried out in a regulated manner, compatible with nationally and internationally agreed principles, standards and obligations [1–15]. A prerequisite to implementing a proper RWM system is to have in place proper financial arrangements as well as organizational and administrative arrangements that define the requirements, competencies, responsibilities and activities of the institutions involved [6]. A key goal in the management of radioactive waste that cannot be cleared, discharged, recycled or reused, is to process the waste to a suitable state for disposal (or storage, if a disposal facility is not available) in order to ultimately safely dispose of the waste. This implies that each waste package has to ultimately comply with the disposal facility WAC (or WAC for the storage facility). If the acceptance requirements for disposal does not exist, WAC need to be developed based on reasonable assumptions about the anticipated disposal route [8].

Since RWM is a stepwise process, typically involving several organizations, in most cases the responsibility for waste passes from one facility or institution to another as waste packages are physically moved from one location to another. Consequently, the liability and risk associated with the wastes are also transferred. To manage these liabilities and risks, WAC need to be defined and implemented for each stage of waste management involving a transfer of responsibility. In this manner, WAC facilitate clear and effective linkages across the interface between the facilities and processes in which material is generated, kept in storage and ultimately disposed of [4]. Therefore, there is a need to clearly define the specific responsibilities of the various organizations involved in waste management, to enable the safe and efficient transfer of waste to the next management step.

Requirement 4 of IAEA Safety Standards Series No. GSR Part 5 Predisposal Management of Radioactive Waste [8] prescribes responsibilities of the operator to ensure an adequate level of protection and safety by various means which depend on the complexity of the operations and the magnitude of the hazards associated with the facility or the activities concerned. Such means include: "Derivation of operational limits, conditions and controls, including waste acceptance criteria, to assist with ensuring that the predisposal radioactive waste management facility is operated in accordance with the safety case". Requirement 3 of IAEA Safety Standards Series No. SSR-5, Disposal of Radioactive Waste [9] also prescribes the responsibilities of the operator and among its provisions states in para. 3.14 that:

"The operator has to establish technical specifications that are justified by safety assessment, to ensure that the disposal facility is developed in accordance with the safety case. This has to include waste acceptance criteria (see Requirement 20) and other controls and limits to be applied during construction, operation and closure."

As indicated in Fig. 1, Section 2.1, and further detailed in Fig. 2, the development process of WAC needs to consistently address interdependencies between the different steps in waste management, starting with the generation of raw waste, and progressing to the completed waste packages that will be disposed of. The disposal facility operator interfaces on a technical level with a predisposal operator primarily through an understanding of the needs and commitments associated with WAC, which are primarily based on the disposal facility safety case. The predisposal operator, in turn, interfaces with waste generators through another set of WAC, which are based on the safety assessment requirements for pretreatment, treatment, conditioning, storage, etc., as well as the disposal facility WAC concerned with disposal. Storage needs to be seen as an additional step that may or may not be required, depending on national circumstances. In some instances, waste can be disposed of directly after treatment and conditioning if a disposal facility is available, while in other cases an interim storage facility may be required until a disposal facility is available.

Considering the various interdependencies in the development of WAC, the radiological, mechanical, physical, chemical and biological properties of waste packages need to be assessed and

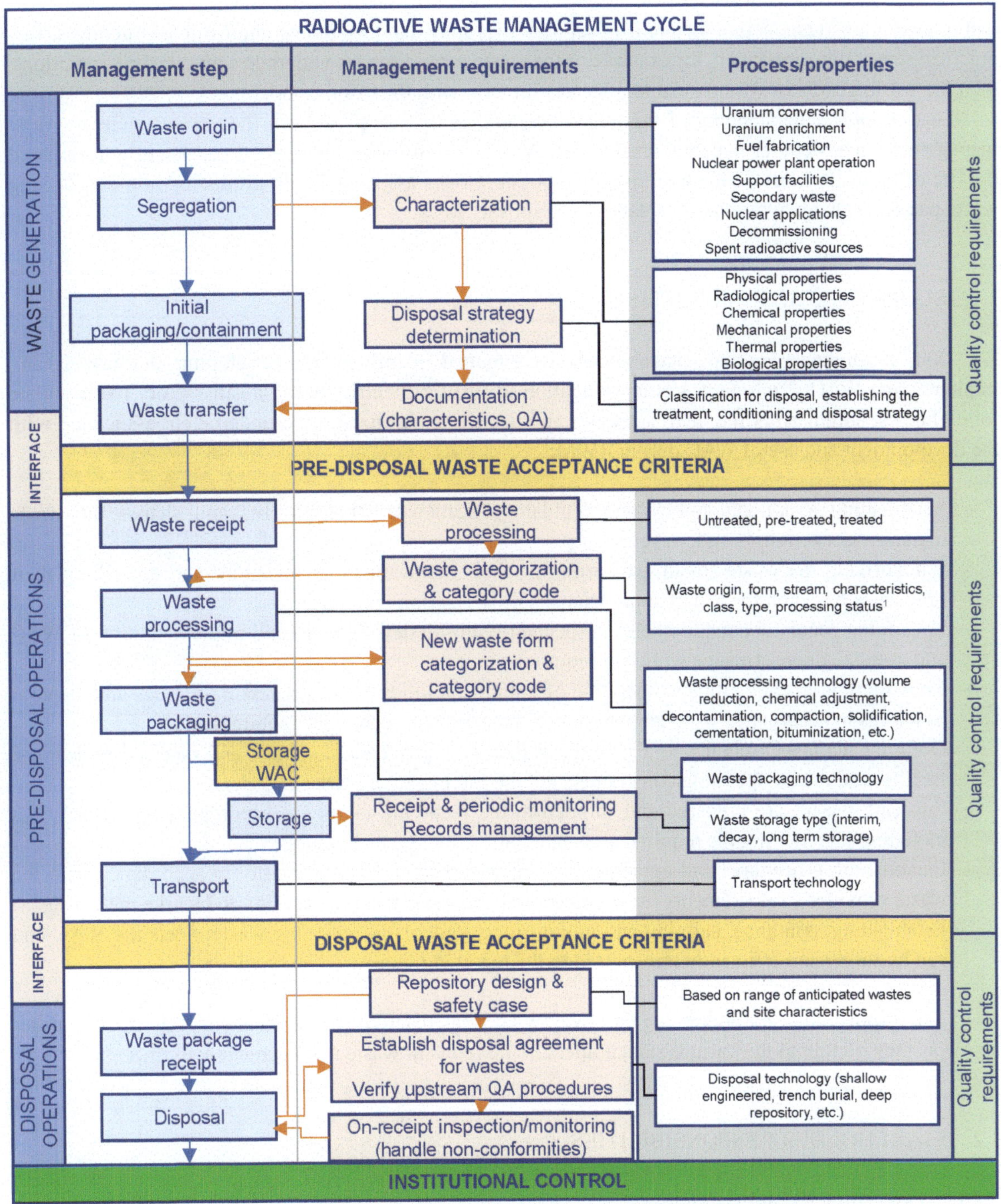

FIG. 2. *The interfaces between different waste management entities. ¹See tables II, III and IV in Ref. [18].*

controlled to be compliant with the transport, storage and disposal requirements. To achieve this objective, the waste package specifications will consider the design and operation of any facility receiving waste and the IAEA's Regulations for the Safe Transport of Radioactive Material [3]. Such specifications will ensure that limits or conditions, which serve as a basis for development of WAC, are addressed appropriately.

It is worth mentioning that in practice, waste management circumstances can vary from case to case such that the distinct division of roles of waste generators, waste processors and disposal operators may or may not be in place. Frequently, waste generators also perform various predisposal/processing operations, and organizations responsible for predisposal activities may also be storage and disposal operators. In a practical sense, this means that, for example, if the waste generator also undertakes predisposal operations, then the waste generator would need to interface directly with the disposal operator.

A thorough understanding of the interdependencies that exist between the various steps in waste management in a Member State will ensure that WAC are developed and applied consistently throughout the life cycle of the waste. This will minimize possible risks associated with producing non-conforming waste packages that may not be acceptable for disposal.

3.1. SCOPE OF RESPONSIBILITIES

Various organizations and stakeholders are engaged in influencing, developing, authorizing and implementing WAC. Each party has an obligation to fulfil a specific role in a satisfactory manner. The following list summarizes the main activities and decisions that need to be undertaken, associated with the development and use of WAC during RWM:

— Establishing a national policy and regulatory framework that governs and shapes the waste management activities undertaken;
— Categorizing the waste streams, determining their inclusion in an inventory and describing their properties based on the use of appropriate characterization methods;
— Specifying regulatory requirements associated with the development and implementation of WAC and assessing conformance with said requirements;
— Constructing safe and efficient operation of suitable facilities for the processing, storage and disposal of wastes in accordance with national policy, strategy and regulatory requirements;
— Defining the WAC relating to a facility, in line with processing, storage and disposal objectives, facility design and operating constraints, and the facility safety case;
— Ensuring wastes are prepared and shipped to the receiving facility in accordance with the WAC, including undertaking the required characterization and documentation;
— Establishing procedures and agreements for waste acceptance at a facility in accordance with the waste provider's requirements for shipment and the waste receiver's ability to receive the waste;
— Establishing assurance, auditing and monitoring protocols for ensuring waste meets the WAC and can be safely and efficiently transported to the receiving facility when required.

The responsibility for addressing these key WAC related activities and decisions is discussed further below as they pertain to the various stakeholders in the relevant waste management activities.

3.2. STAKEHOLDER RESPONSIBILITIES

3.2.1. Governments and regulators

The governments of Member States are responsible for defining a national policy for RWM, ensuring that appropriate legislation is in place and establishing a regulatory framework. Each Member State is also responsible for developing and implementing a national RWM strategy. The national programme is defined by the alignment of policy and strategy, legislation and applicable regulations [6].

The national regulatory body, and local regulatory bodies where they exist (e.g. state authorities in a federal system), are responsible for the licensing (including WAC approval) and monitoring of various RWM facilities and practices, as indicated in Fig. 3. Some regulatory authorities may be responsible for the development of WAC as well as for ensuring adherence to them. It is also noteworthy that different regulatory authorities in a Member State may be responsible for different areas of a RWM programme (e.g. concerning radiological and environmental aspects).

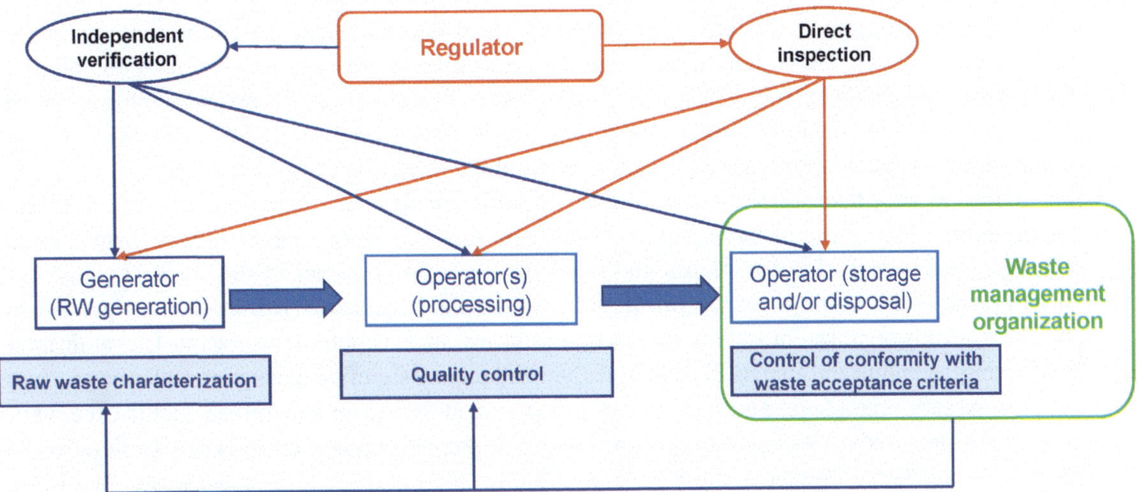

FIG. 3. General scheme showing the responsibility of the regulator between the different waste management entities, from raw waste (RW) generation through storage and/or disposal.

Every country needs to clearly define its regulatory framework, which could include the specification, acceptance, concurrence and approval of WAC, as well as ensuring implementation and conformance. A key component of regulator expectations is the use of well defined WAC in the preparation of waste forms for storage and disposal [10, 16].

Examples of Member States where the responsibilities of the government and regulators are clearly established and where there is a well functioning relationship between the government, regulator and waste operator are given in Annexes I to VI for Belgium, France, Slovakia, Spain, Japan and Hungary.

3.2.2. Waste generators

A waste generator is responsible for the safe management of the waste it produces in accordance with government policy and regulatory requirements and to the satisfaction of the public and other stakeholders. This requires the establishment and implementation of an appropriate, safe and compliant strategy for both the generation and management of waste. The waste generator will need to liaise with the operators of waste processing or disposal facilities regarding their future receipt of waste, addressing especially the following:

— The origin and timing of anticipated waste streams, degree of waste form segregation and characterization methods;
— The volume and nature of the anticipated inventory;
— Contributions to the development of WAC by providing relevant information about the waste and discussion concerning their interpretation;
— Procedures, methods and resources to be adopted to ensure compliance with the WAC.

Thereafter, the waste generator is responsible for the preparation of the wastes for shipment and transfer to the receiving facility in conformity with the WAC. To fulfil this responsibility, the waste generator will ensure that the following waste management aspects are addressed:

— The availability of adequate resources.
— The establishment and operation of a facility and the use of equipment required for preparing the waste for shipment to another facility.
— The provision of an accurate and comprehensive waste inventory. This will include a record of existing wastes and projections of future wastes that are expected to be generated, so that the

13

waste receiver's requirements may be fully incorporated into the waste generator's internal waste management strategy. The comprehensive waste inventory keeps the waste receiver fully appraised of the quantity and characteristics of the wastes to be received. Reference [19] provides information on developing and implementing a methodology to retrieve, assess, verify and restore lost or inadequate waste inventory records for legacy storage or disposal facilities (historical waste).

— The ability to undertake waste characterization of sufficient accuracy and precision. This is crucial for the further processing of the waste at downstream facilities since characterization begins at the waste generator's site. The techniques and equipment used to characterize the waste streams need to sufficiently address physical, chemical and radiological information requirements stipulated by the downstream organizations. This may require the use of a variety of assessment, examination, monitoring, sampling and analytical techniques and will require significant planning and infrastructure.

— After necessary segregation and preliminary processing at the waste generator's facility, the waste needs to be prepared for transport to the next facility. Transport requirements [3] may be required for shipments depending on which entity is responsible for the transportation of the waste; that is, the generator, the receiver or a third party.

— A quality assurance/control system and associated procedures (addressing technical, organizational and administrative measures) to ensure that the WAC are met to the satisfaction of the waste receiver, will be required. Ensuring that an adequate system is established will require regular liaison and agreement between the waste generator and waste receiver. Interactions could include permitting audits undertaken by the regulator.

— Preparation of a data file for each waste package and completion of the waste receivers' documentation according to their requirements.

— Physical transfer of waste to the receiving facility along with appropriate documentation.

In the UK, generators of radioactive waste report anticipated waste arising and agree a management strategy with the Nuclear Decommissioning Authority (NDA), which maintains a detailed national waste inventory that addresses radionuclide content, physical form and specific conditioning and disposal strategy. In agreement with the NDA, waste generators ensure LLW meets the relevant WAC for each waste stream and is disposed of appropriately; this is done through consultation with the national LLW disposal facility. ILW is conditioned typically by cementation into a passively safe form and held in engineered storage facilities. WAC for conditioning and storage are derived from WAC developed for a conceptual deep geological disposal facility. ILW is stored in packages that are suitable for long term storage until such a facility is available. Examples from several Member States of the key responsibilities of waste generators are given in Annexes I–VI.

3.2.3. Waste receivers

The waste receiver may comprise:

— A waste processing facility (e.g. a super-compaction, metal melting or immobilization facility);
— A storage facility; or
— A disposal facility.

Separate organizations may be responsible for each type of facility in a Member State, or one organization may be vertically integrated and responsible for two or more steps. In principle, waste receivers are responsible for the following tasks:

— Understanding the nature and quantity of waste to be received and accepted.
— Designing, constructing and operating a facility in accordance with its safety case to produce the treated/conditioned waste product or to dispose of the waste in accordance with the planned intent.

— Liaising with the waste generator and developing accurate and comprehensive WAC that are understandable and achievable.
— Ensuring that regulatory authorities are aware of the WAC and authorizing their use as necessary.[1]
— Disseminating WAC to ensure that wastes will be received in a form and state that is compatible with design and safety case requirements (including transport).
— Monitoring, reviewing, updating and improving WAC to ensure they remain fit for purpose.
— Establishing a procedure and process for agreeing the receipt of wastes according to the WAC. The agreement process will include qualification of the waste production route and verification that quality assurance/control procedures are being applied, as well as administrative and physical checks of waste packages.
— Receipt of wastes on shipment to the facility and taking title for those wastes in order to transfer liability.
— In the absence of a disposal facility, liaise with the storage or processing facility to ensure that waste treatment processes will produce waste packages in accordance with any proposed or generic WAC for disposal if such exist, as far as is practicable by not foreclosing options while maintaining safety and security.
— Maintaining and making available full records of the waste for future waste management purposes.

The following sections outline specific responsibilities of the different types of waste receiver: waste processors, storage facilities or disposal facility operators.

3.2.3.1. *Waste processors*

Processing is any operation that changes the characteristics of waste and includes pretreatment, treatment and conditioning [15]. The waste processor will implement an appropriate management system that includes quality assurance and control in a similar manner to the system of the waste generator. This is required for demonstrating not only the safety and quality of the work being undertaken, but also the fulfilment of the relevant WAC stipulated by the next responsible institution in the waste life cycle. Waste processors will ensure that the characterization of the modified waste form and final waste packages conform to the disposal WAC, and they will provide all necessary information for the continued safe management of the waste. This is particularly important if the treatment substantially changes the characteristics of the raw waste, such as by causing a significant modification of its chemical or physical form, waste sorting or blending. Examples would include super-compaction and overpacking of pucks in a drum (waste mixing), dewatering of sludge wastes (volume reduction, loss of dissolved chemical salt content), pyrolysis/ashing of organic wastes and immobilization of the residue (substantial chemical change and volume reduction, with potential loss of volatile radioactive components), cementation, vitrification and plasma melting. Historical data pertaining to the generation of a waste package will be included as appropriate.

3.2.3.2. *Storage operators*

A dedicated waste storage facility would typically be used for interim storage of raw packaged or loose waste pending further treatment or conditioning or for medium to long term storage of conditioned waste pending transfer to a separate facility for disposal. The storage facility will apply WAC related to its own safety case, and design and operational configuration for the handling, placement and monitoring of wastes. However, the characteristics of the waste packages that are received for storage, and the arrangement of the store and handling facilities, will need to anticipate and reflect the next waste management step. This will be particularly important if the wastes will be routed to disposal, since by design there may be no additional conditioning of the waste package for disposal. Therefore, WAC for storage will largely correlate with the WAC of the designated disposal facility, with additional criteria for storage related to its own safety case, facility design and operation, environmental conditions and any additional safety requirements. The storage operator will be expected to demonstrate the fulfilment of the waste destination WAC to the satisfaction of the next responsible operator and ensure that the WAC applied for the receipt of the waste are fully compatible with those of the next institution unless, for example, there is re-packaging or overpacking at the storage facility, so that the onward transport of waste and the transfer of liability can be successfully undertaken.

An important task for the store operator, particularly for long term storage, is the maintenance of comprehensive waste records such that full characterization information, together with waste condition monitoring information obtained during the storage period, may be passed to the disposal site operator.

3.2.3.3. *Disposal operators*

The disposal facility operator is responsible for the design, construction, operation and closure arrangements, which need to comply with national regulatory requirements, particularly with respect to the overall safety of the disposal facility during both operational and post-closure periods. The disposal operator is responsible for specifying and implementing appropriate WAC to satisfy these requirements. Starting from generic or preliminary requirements and following an iterative process, the operator needs to revise and refine these criteria to address developments in the realization of a national waste management strategy through the implementation of a disposal route. This work will be performed in close cooperation with the waste generators and processors, storage operators, regulators and any other institutions or organizations involved in the management of waste intended for disposal. A comprehensive list of disposal operator responsibilities is provided in Section 3.2.3 on waste receivers.

Additionally, the disposal operator could provide direction for and oversight of activities relating to routeing the waste for storage and/or disposal or the characterization and conditioning of raw and treated waste.

A key aspect of the disposal site operator's responsibility for WAC specification is addressing issues that affect post-closure disposal facility safety such as waste form stability, leach rates and information on any long-lived radionuclides present in the waste that would potentially give rise to a dose detriment after disposal. Aspects applicable to short term management such as dose rate limitation and handling arrangements are included in the WAC for predisposal operations and storage, if relevant.

Some of the long-lived radionuclides important for a disposal facility safety case are not easily identified in the wastes, e.g. alpha-emitting or non-gamma-emitting radionuclides such as ^{14}C, ^{36}Cl, ^{59}Ni, ^{99}Tc, ^{129}I, ^{135}Cs, etc. The disposal operator will collaborate with the waste generators and processors who need to prove that they have adequate means and capability to reliably determine the content of such nuclides when setting the acceptance criteria. The long term detriment of chemically toxic species will need to be considered when specifying acceptance criteria for chemically hazardous species.

3.2.4. Waste transporters

Note that all transport operators will have to comply with the internationally and nationally accepted transportation regulations for the movement of radioactive materials [3]. Such regulations are integrated in the transport requirements and regulations.

3.2.5. Third parties

In some cases, third parties (providing an independent service to a facility operator or regulator) could be tasked or contracted to certify or undertake certain steps in the waste management system, for example, the transportation of radioactive waste. However, the primary responsibility for the waste always remains with the customer commissioning the service. If services are outsourced to a third party, the management system of that party needs to be documented, particularly with regards to quality assurance and control, with qualification and audits undertaken as appropriate.

3.3. ESTABLISHING THE WASTE ACCEPTANCE CRITERIA

According to GSR Part 5 [8], an operating organization is responsible for the development of operational limits and conditions and controls, including WAC. WAC are to be specific to a waste management facility and based on the operational and post-closure safety assessments. WAC documentation is generally included in the safety case for a facility. The safety case, incorporating safety assessments and WAC amongst other evidence and argumentation, is to be submitted for approval by an appropriate regulatory body [10]. In addition, it is the operator's responsibility to ensure that sufficient information about the waste under its remit has been collected and recorded at each prior step in the RWM process in order to demonstrate compliance with downstream WAC (most often included in the safety case) [10].

The responsibility for the preparation, acceptance, implementation, maintenance and revision of WAC documentation will vary depending on the regulatory structure of the Member State. In many cases, the facility operator maintains documentation to control waste coming into the facility and to ensure compliance with the operating licence. In the case of a disposal operator, the documentation ensures an acceptable level of post-closure safety. As such, it is prepared by the facility operator (with approval of the regulator) and issued as contractual conditions that will be strictly followed by the sending organization. In some cases, the document might be a regulatory instrument issued to the facility operator, who is then responsible for ensuring compliance.

When defining WAC, operators need to consider the implications and consequences for the operation of all facilities involved in the waste management system. It is important to note that any plan for RWM will provide, as far as practicable, a clear indication of the waste types, waste classes, waste quantities and nuclide inventories of the waste to be managed and disposed of.

With respect to possible deviations, an operator will have to implement appropriate measures and procedures for dealing with departures and non-conformities. Departures need to be identified before waste packages are produced. Non-conformities can only be identified after the production of waste packages and during checks, unless declared by the originator on the basis of prior knowledge. Departures and non-conformities can be treated correctly only when all the necessary information and documentation are available.

Arrangements regarding the transfer and ownership of waste and waste packages between waste management facilities, as well as resulting nuclear liabilities, will be agreed upon and documented as noted above. Such agreements are usually contained in the form of formal contracts and may include payment schedules as well as responsibilities and liabilities for addressing non-conforming waste and waste packages.

3.4. PREPARING AND MANAGING WASTE RECORDS

Proper documentation and record keeping are fundamental components of a management system and necessary elements of RWM. The necessary documentation includes information on the radioactive inventory, package identification and characterization records, package shipment and receipt dates and storage and emplacement locations, monitoring records, etc. Waste records are essential in assessing the compliance of disposed wastes with current safety and QMS requirements, and compliance with revised WAC for long term behaviour modelling calculations.

Generally, the more data obtained in the early stages of a waste life cycle the more straightforward future RWM will be. It is important to establish early on a suitable record keeping system compatible with systems elsewhere in the chain of RWM and with longevity in mind.

Records on the management of radioactive waste will be maintained throughout the management process, ensuring adherence to strict quality controls and compliance with relevant regulatory requirements. Table 2 shows the minimum content of records generated by facility type, which need to be established and retained throughout the waste management cycle [20].

It is the waste receiver's responsibility to clearly specify documentation and record keeping requirements as part of the WAC specification, and the waste sender will have to anticipate and fully adhere to these requirements and provide all information that could be deemed of value in future waste management activities.

TABLE 2. MINIMUM CONTENT OF RECORDS IN RELATION TO WASTE PROCESSING, STORAGE AND DISPOSAL

Organization	Required record content
Generator	— Waste generating process and pretreatment (e.g. collection, segregation); — Volume and mass of waste of waste streams and dates of arising; — Assay reports; — Analytical characterization (chemical, radionuclide composition and activity) of raw waste, characterization methods; — Inferred radionuclide content (e.g. from scaling factors); — Documentation concerning preparation, packaging and dispatch of raw and/or pretreated waste, addressing downstream WAC requirements.
Processor	— Inspection and administrative checks to confirm waste receipt and conformance with WAC and traceability from the waste generator documentation up to final package (or batch) identification; — Waste processing records including information related to the critical processing parameters; — Assay reports; — Analytical characterization (chemical and radionuclide composition) of conditioned waste; — All waste package characteristics required by downstream WAC (inspections, shielding, container configuration, etc.); — Independent inspection reports (where required); — Non-conformity records.
Storage operator	— Inspection and administrative checks to confirm waste package receipt and conformance with WAC, including container identification with a link back to waste package records; — Individual package location; — Reports of periodic inspections, monitoring plan implementation and remedial actions; — Details related to the storage conditions; — Non-conformity records.
Transporter	— Container identification with a link back to waste package records; — Waste package acceptance for transport; — Demonstration of compliance with transport regulations; — Timing of dispatch and delivery and routeing; — Incidents or accidents during transport.
Disposal operator	— Inspection, analysis and administrative checks to confirm waste package receipt and conformance with WAC, including container identification with a link back to waste package records; — Emplacement location and date; — Non-conformity records including final resolution.

The documented receipt and transmittal of records at each step of the waste life cycle is advisable to be established in specific procedures. Upon receipt and based on a documented system that provides for the ease of retrieval and maintenance of traceability of records to their associated waste process and/or product, the records need to be indexed. When different organizations are involved in the incremental waste processes, document management system coordination among all organizations is essential in facilitating the record use and retrieval.

The proper control of the records distribution to authorized organizations and/or designated retention locations needs to be in place, and the record content will be established and approved for each step in the life cycle of the waste.

The information identified in Table 2 is extracted from the records and will serve as a basis (in addition to the basic waste characterization) for further technological decisions; these records are in most cases transmitted and used by the downstream organizations. Briefly, this can be exemplified as follows: the conditioning facility is receiving from the waste generator a summary of the process knowledge relating to individual items of as-generated waste and is further using this information to establish an appropriate treatment and conditioning pathway. The conditioning facility will retain the process knowledge documentation (to be used in future investigations or audit processes) and will generate

TABLE 3. INTERACTIONS AND RESPONSIBILITIES OF DIFFERENT ENTITIES RELATED TO WASTE RECORD GENERATION, DISTRIBUTION AND RETENTION

Type of data	Generator	Processor(s)	Storage operator	Transporter	Disposal operator
Raw waste — physical and chemical composition	a	b			c
Raw waste — radionuclide inventory	a	b			c
Raw waste — activity measurement	a	b			c
Physical and chemical composition of conditioned waste		a	c	c	d
Container characteristics		a	c	c	d
Process and matrix characteristics		a	c	c	d
Waste package characteristics (e.g. waste form, weight, radiological inventory, dose rate, surface contamination)		a	c	c	d
Location of waste package in the storage facility			a		
Waste package monitoring details during storage			a		d
Proof of compliance of waste package with the transport regulations requirements				a	
Location of waste package in the disposal facility					a, d
Records of independent inspections (if requested)					a, d
Records related to the non-conformities and any other operations performed	a	a	a		a, d

[a] Responsible for data generation.
[b] Responsible for records on information related to waste processing.
[c] Package traceable to records.
[d] Responsible for records containing relevant information for the post-closure phase.

the characterization files (records) for the related conditioned waste forms. These records will further facilitate the compliance verification of the waste package with the storage or disposal facility WAC. The conditioning records will need to be traceable by package identification number to allow the in time action in case of unwanted events such as a content release or container integrity failure. If a disposal facility or concept is not available, and the waste needs to be stored for longer time frames, it may be appropriate to transfer the complete documentation to the storage facility operator. For transport purposes, the necessary documentation is limited to a summary of package characteristics to evaluate the compliance with the transport requirements.

Table 3 provides the general logic of interaction between different entities and shows distinctly defined responsibilities assumed for the waste generator, processors, storage operators, transporters and disposal operators for the generation and retention of records supporting the acceptance of waste packages.

3.5. USE OF QUALITY MANAGEMENT STANDARDS

Implementation of an appropriate management system throughout the waste management process, including a QMS, is essential to ensure traceability, consistency and confidence in the waste management process.

Waste management facilities will have to demonstrate compliance with a quality management standard that will ensure efficient control and operation of the waste management processes through periodic checks and audits. This may include:

— Internal quality assurance (QA) approval where objective measurements (audits) are performed by a company or a department upon its own systems, procedures and facilities. The auditors may be from the company's own ranks or from an external environment.
— External QA approval where an accreditation body conducts audits as a result of successful registration to an accreditation scheme.

In this regard, the ISO 9001 international standard [21] is widely used because it deals with QMSs for both internal and external quality management processes and represents current best practice in the nuclear and non-nuclear industries. It prescribes requirements for key areas such as management responsibility, control of documents and records, design and development, identification and traceability, inspection, measuring and testing, continual improvement, corrective and preventive action and independent verification.

Where different standards for quality management are used, a clear framework that indicates the intended use of the standards as well as their relevance to WAC needs to be provided.

All other relevant legal and regulatory standards will be considered during the different stages of the waste management process, including those specified in laws, standards, legal directives and codes.

4. DEVELOPING WASTE ACCEPTANCE CRITERIA

4.1. ITERATIVE DEVELOPMENT

As was highlighted in Section 3, interdependences exist between all steps in the management of radioactive waste, and therefore planning for all the various steps needs to be performed in advance so that a balanced and coordinated approach to safety is taken. Advance planning will avoid any potential conflicts between the safety requirements and the technical and technological requirements in the overall waste management programme. In particular, the following topics need to be addressed:

(a) The identification and documentation of interfaces between each waste management step, clearly defining the responsibilities among the involved organizations;

(b) The establishment and conformance with the WAC for each waste management facility [8].

WAC are developed for each step in the waste management process and are used to facilitate the safe and efficient physical transfer of waste and associated liabilities from one organization or facility to another, as has been discussed above, as well as to assure the safety of radioactive waste once received at a processing, storage or disposal facility. The development of WAC for an interim step (e.g. processing or storage) needs to consider the final step in the process (e.g. disposal) as discussed, in order to avoid extensive re-working of the waste package to meet consequent WAC [11]. The WAC for later steps needs to be considered as much as possible when defining WAC in earlier steps. The premise of this integrated approach is that the majority of the waste can therefore pass from one waste management step to the next as a matter of routine. Only a small portion (if any) of the waste would need to undergo re-examination between the steps if there is confidence in the waste acceptance process, as demonstrated by experience involving audits and other checks and the adoption of a robust management system. Figure 4 indicates the relationships between key influencing factors in development of WAC to be performed based on the safety assessment results within the safety case for a RWM facility according to Refs [10, 12].

Figure 4 shows how the decisions related to the predisposal management of radioactive waste need to be taken and implemented to ensure compliance with the WAC for the available or anticipated disposal options [11]. However, for many RWM programmes, important decisions have still to be made before the WAC for disposal can be finalized, because, for example, the disposal facility does not yet exist.

The waste form qualification process is intended to collect evidence that a particular waste form and package can be accepted by storage and disposal facilities without compromising their operational and long term safety. The process consists of extensive studies on the performance of a waste form under disposal facility conditions, identification of destruction scenarios, mechanisms and pathways that allow for the quantification of radionuclide release. Typically, the waste form will need to be subjected to the evaluation of long term leaching and water immersion, gas generation, porosity and diffusivity, freeze–thaw cycling, compressive strength tests, resistance to microbial, radiation and thermal effects, chemical performance and influence of chemical species present in waste, etc. Some of the tested parameters can be selected to become WAC (e.g. leachability, compressive strength).

In this section, the discussion mainly focuses on the development of WAC for a disposal facility, on the basis that ultimately disposal WAC will influence the development of predisposal WAC and that

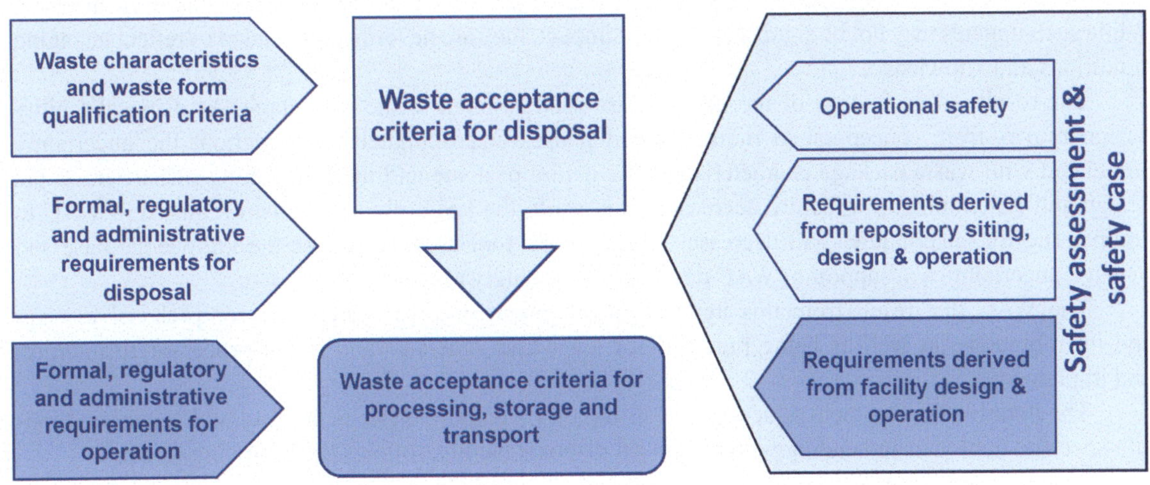

FIG. 4. *Relationships between requirements and influencing factors in the development of disposal and predisposal waste acceptance criteria.*

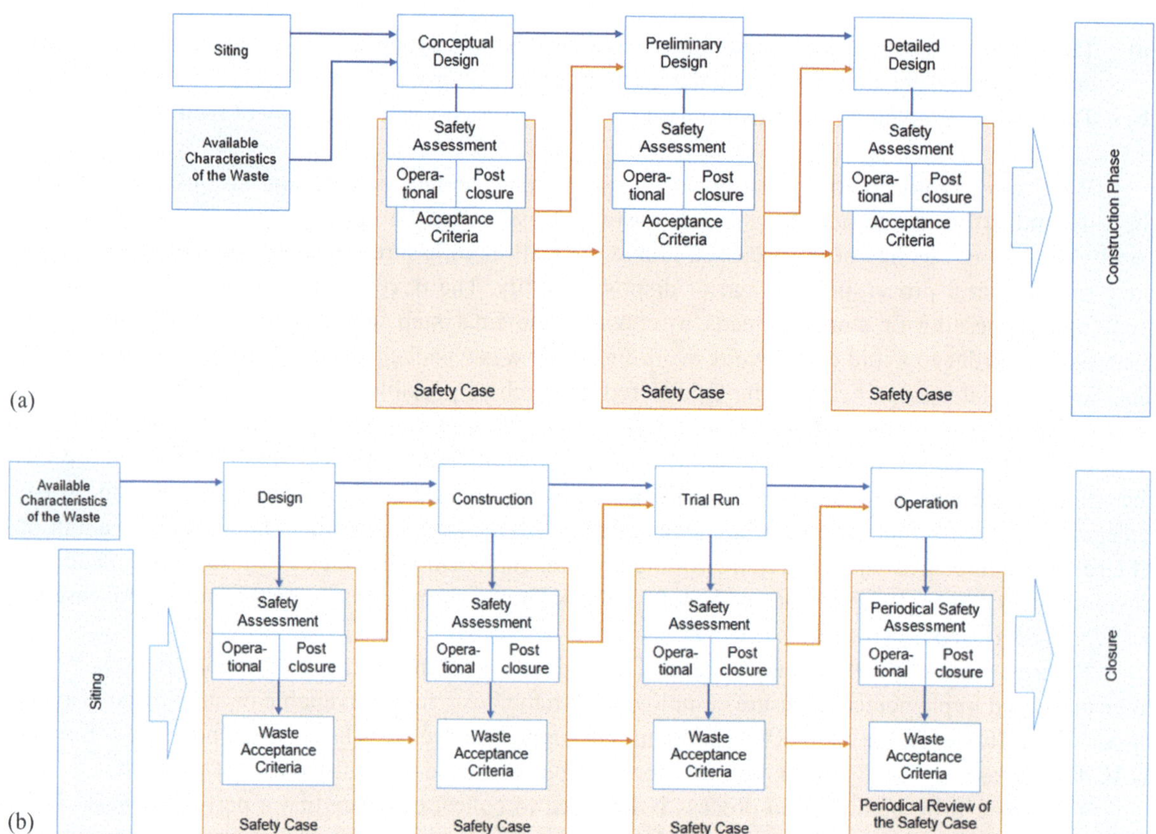

FIG. 5. Typical iterative process for the development of disposal facility waste acceptance criteria (a) during the design phase and (b) from design to the operational stage.

the majority of principles relating to the development of disposal WAC will have application in upstream processes and facilities. Comments will be made where appropriate, to justify the application of the following information for predisposal processes.

As alluded to earlier and as shown in Fig. 5, the development of WAC for disposal is an iterative process and needs to be conducted to reflect all updates and changes in the safety assessment and the safety case, and the design of the disposal facility. This process needs to be viewed as a continuing cycle because the WAC are likely to be further updated even after the disposal site has been licensed. While such updates may not be significant in their impact, they are nevertheless needed to reflect changing conditions and knowledge.

The results of each stage of the iteration are used as input to the next stage. As disposal facility design moves from conceptual to more detailed phases and ultimately construction, the uncertainty associated with waste package characteristics, the nature of a site and the functioning of barriers in the design and functioning of a facility decrease. As a result, the level of conservatism required in WAC to compensate for uncertainties will decrease. The principal tool used to analyse the knowledge base and manage uncertainties in support of WAC development is safety assessment.

The WAC that result from this iterative development process need to be comprehensive, clear and unambiguous, as well as being practical for the waste generator and waste processor to interpret and implement.

The iterative development of acceptance criteria for a disposal system, starting with a generic (non-site-specific) approach and ending with a licensed disposal facility, follows the following steps:

(a) At the initial disposal facility planning stage, preliminary WAC are defined in a very general form based on the existing safety case and accompanying safety assessment which takes into consideration the overall waste disposal strategy and general information on the types and quantities of wastes known or anticipated to be received, both existing and to be generated, as well as a conceptual design for the disposal facility. It will be essential that regulators are involved in discussion at an early stage, to ensure that legal and regulatory requirements and expectations are understood and possible to meet. This involvement may be formal or informal, depending on national circumstances. Comparable industry experience, including international best practices, may be used in developing preliminary safety assessments as a basis to further develop qualitative generic WAC, which are applicable to any facility, and this includes:

(i)   Administrative requirements (contractual issues, data packs, certificates, records, etc.);

(ii) Waste form requirements (prohibited waste, explosive materials, toxic materials, corrosive materials, biological pathogens, etc.);

(iii) Quality requirements (inspection, testing, verification, etc.).

Such preliminary safety assessments would incorporate information regarding the waste inventory for disposal, the characteristics of potential disposal sites and possible disposal configuration options within a conceptual design. Safety assessment is a component of the safety case, which specifies the measures that need to be taken to achieve the protection objective of disposal and defines the principles which are applied to demonstrate that this objective will be reached. The safety case emphasizes the importance of site selection, the multibarrier concept and the use of state-of-the-art technology, among other factors.

(b)   After site selection has commenced and the general characteristics of potential disposal sites become known, a more detailed safety assessment of the disposal system can be made using site specific data and information, albeit provisional. At this stage an understanding of the relationship between waste package characteristics/parameters, the relevant safety case and accompanying safety assessment, site conditions and disposal facility design may lead to the formulation of preliminary quantitative WAC (see Annex VII for more details). Requirements for disposal will indicate, through safety assessment, relevant waste package parameters that need to be incorporated into WAC. Additional requirements impacting WAC may be imposed by the licensing and supervising authorities or from practical considerations. Individual requirements may apply to a broad range of different types of waste packages or individual types of waste packages. Naturally, there are often options available, especially concerning disposal facility design. This allows some degree of freedom of judgement and approaches in the early development of WAC.

(c)   Once the suitability of a site has been confirmed and the final characteristics of the disposal system established (that is, the disposal facility has been designed in detail), a safety assessment undertaken prior to authorization and construction leads to the definition and stipulation of the quantitative and qualitative WAC.

(d)   Once the site has been selected, the facility designed and constructed and active and non-active trial run performed, the safety assessment will be updated and the final version of the safety case for operation compiled. Such a safety case and accompanying safety assessment will consider all changes introduced since the initial safety case and safety assessment. Based on this safety case, the WAC which will be used during operation of the facility will be established. After formal approval by the regulator to use and apply these WAC, the disposal facility can begin to accept waste and the programme moves into the operational phase.

From the above, it is clear that in order to achieve the disposal protection objective, an iterative process need to be in place. The process aim is to bring together detailed information as the disposal facility project progresses through for each phase of investigation, conceptual planning, detailed design and performance assessment. Furthermore, the preliminary WAC (generally presented as guidelines initially) develop based on the safety case and accompanying safety assessment taking into account its evolution thus becoming more advanced WAC, many of which are quantitative, that are suitable for operations.

After licensing and the start of disposal operations, WAC may be further modified to reflect operating experience, technical development and scientific progress, etc. This approach is illustrated in

Fig. 6, which more clearly sets out some of the important components influencing the development of WAC at different stages in a disposal programme.

Key uncertainties that ultimately impact on dose and risk gradually diminish throughout the development of a RWM programme, as more data become available and key decisions are made. To determine the significance of new data, information and understanding, a site specific safety assessment that includes an evaluation of the required safety of the disposal facility in its operational and post-closure

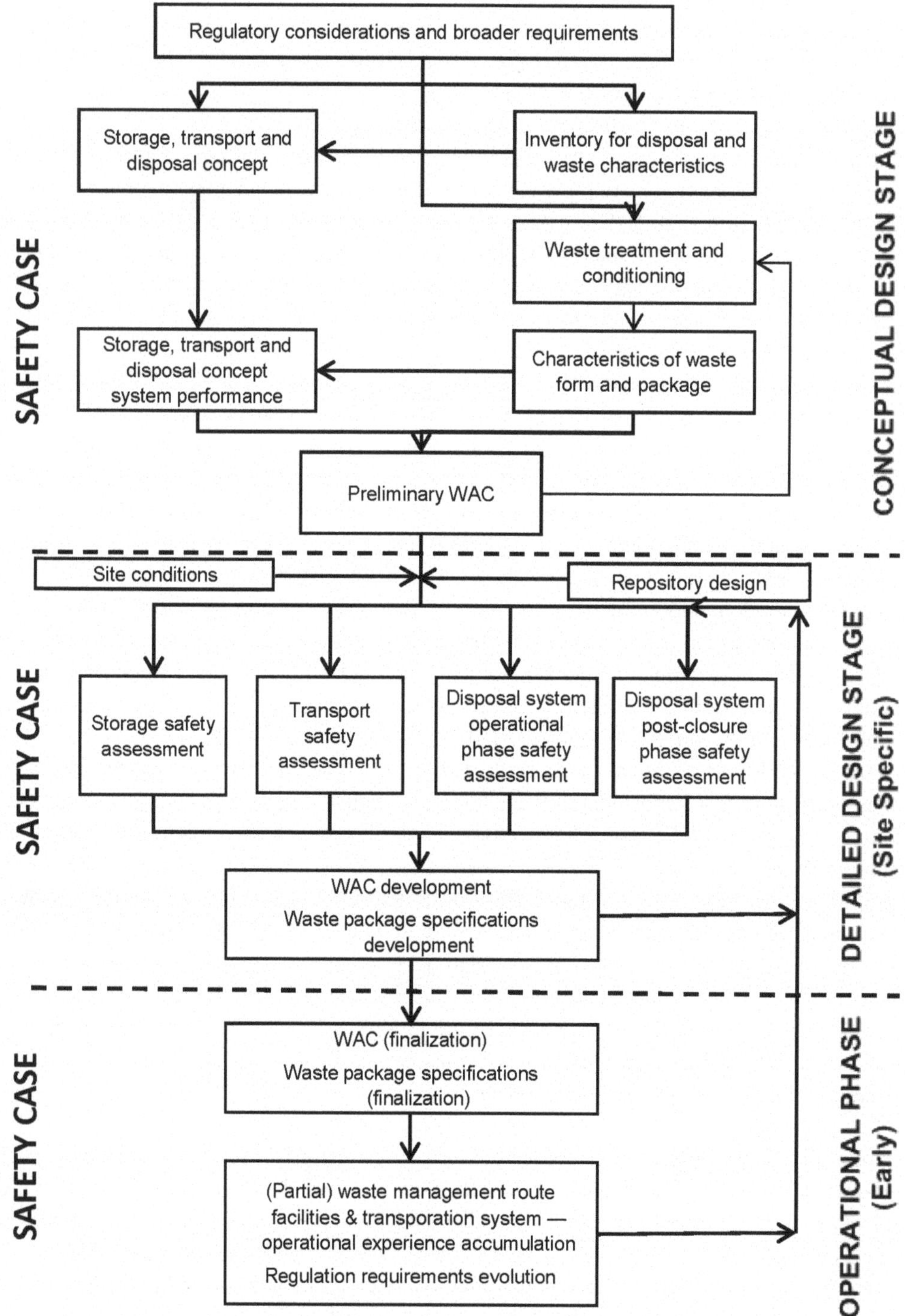

FIG. 6. Stages in the development of waste acceptance criteria (WAC) for disposal.

phases needs to be updated. This in turn supports the derivation of requirements placed on the nature of the waste packages destined for disposal.

The preliminary waste acceptance requirements (guidelines) thereby develop to become quantitative WAC. The development of the quantitative waste acceptance requirements for a disposal facility are relying on the results of the site specific safety assessment which consider several aspects, such as the geological situation, the technical concept or design of the disposal facility, the mode of operation, and the waste package characteristics [22].

It is important to highlight that this iterative approach is also followed in the development of WAC for predisposal steps. For storage of waste packages, the criteria are also developed through an iterative process between package characteristics and storage facility design, moving from conceptual to the final design stage. For the treatment and conditioning of raw waste, the iteration is between waste characteristics and considerations related to the nature of the process (e.g. incineration, compaction, ion exchange) and the design of processing equipment. Similar principles apply in the development from qualitative, generic criteria to specific, quantitative criteria.

4.1.1. Nature and level of technical requirements for disposal phases

Operational and post-closure disposal phases bring different technical input data for development of WAC. These can be further divided into two categories: generic requirements and specific requirements, depending on the availability of information according to the progress of technical studies relating to both the waste packages and facilities (at the successive stages of feasibility study/conceptual design, basic/preliminary design, detailed design and so on). This approach is illustrated in Table 4.

Limits of specific and total disposed of activity of radionuclides are dependent on parameters of a disposal facility (design, site characteristics); they are set as a result of the facility safety assessment. Other

TABLE 4. AN OUTLINE APPROACH FOR ESTABLISHING WASTE ACCEPTANCE CRITERIA FOR DISPOSAL

	Requirements in each phase of disposal	
Nature and level of the requirements	Operational phase requirements: Safety requirements that need to be satisfied over the operational phase, covering waste package emplacement and concluding with completion of backfilling	Long term safety requirements: Safety requirements that need to remain satisfied over the long term following closure of the disposal facility
Generic requirements: Safety requirements that need to be satisfied by waste packages and disposal facilities when generic disposal conditions and the general properties of the wastes are assumed	General requirements: • Waste package: Durability, measures to prevent/control leakage of nuclides, etc. • Disposal facility: Structural stability, measures to prevent/control leakage of nuclides, dose constraints, etc. • Activity classification: Classification activity of ILW and LLW, etc. [11]	Additional requirements: • Design target of waste package: Nuclide adsorption properties, long term stability, etc. • Design targets of disposal facility: Dose constraints, exemption level, etc.
Specific requirements: Safety requirements that need to be satisfied when detailed requirements resulting from the detailed disposal facility design and safety assessment are specified	• Design specifications of waste package • Dose rate, weight, long term stability, etc. • Design specifications of disposal facility • Shielding, handling equipment, decomposition/corrosion allowance, disposal facility specification, etc. • Safety assessment Environmental conditions at the disposal site, maximum activity concentrations, performance of barrier, etc.	

parameters, such as administrative and technical, do not require information about the disposal system; furthermore, transport and storage requirements cover most considered waste package characteristics.

4.1.2. Requirements associated with different disposal phases

In the development of WAC for disposal of radioactive wastes, it is important to consider that the safety functions assigned to waste packages differ for operational and post-closure phases. This is because disposal facilities, unlike other nuclear facilities, have the requirement for ensuring safety over a very long period that may extend beyond tens of thousands of years. It is even more important to consider that activities during operational and post-closure periods as well as hazards posed by those activities are different. The timescale for consideration will be decided to reflect the hazard posed by the waste to be disposed of. Over such long timescales, the physical and chemical barrier functions that might be an inherent feature of a waste package would be expected to break down, due to corrosion of metals and degradation of cement, for example. Within the appropriate design, the role of safety assessment and the safety case is to demonstrate with sufficient confidence that any long-lived radionuclides that might be released from a disposal facility would not adversely impact humans or the environment. Therefore, the deterioration of barrier functions, including those associated with a waste package, are not to adversely impact dose and risk. With such assurance, it is not necessary (or even possible) that the initially required performance of waste packages that have been disposed of be maintained indefinitely. This disposal situation is compared with storage, where the requirements are consistent over time in order to maintain the performance of the packages until at least the end of the storage life.

In establishing criteria for waste acceptance, both operational safety and post-closure safety (see Table 4) need to be addressed, as follows:

(a) *Requirements derived from operational safety* — relevant for ensuring safety over the period during which waste packages are emplaced in the disposal facility and until closure (e.g. backfilling and sealing) is completed and mainly relates to requirements to ensure the safety of the workers at the facility.

(b) *Requirements derived from post-closure safety* — requirements for ensuring safety following completion of backfilling/closure, that is requirements pertaining to the post-closure stage. These requirements address the deterioration of waste packages and the disposal structures, and the potential for nuclide migration back to the biosphere as well as the consequences of inadvertent human intrusion when active control of a site is terminated or lost. These requirements mainly deal with the long term safety of the public and protection of the environment.

4.2. DEVELOPMENT APPROACH

A systematic approach that takes into consideration the functions of the different components of the disposal facility — the waste form, the waste container, engineered and natural barriers that are present (barriers whose functions are related to the disposal facility or storage facility site conditions) is followed when developing the safety case which then serves as a basis for the development of operational limits and conditions and WAC. It is important to understand these functions in a holistic manner for the specific waste management system that is considered, to allow for waste management process optimization. Failure to do so could lead to overly conservative, irrelevant or impractical to implement WAC. It is desirable to establish full dialogue between waste generators and processors and storage and disposal operators on waste form, suitable package types, pretreatment requirements and available technology (e.g. sorting, decontamination), viable conditioning methodologies, etc. It is possible that the development of waste criteria specifications will lead to research and development (R&D) concerning waste treatment and conditioning approaches so that the associated acceptance criteria may be properly met for given waste streams. This, itself, is a positive fact as technology advancement will generally be the result.

However, the time and expenditure implications of driving an R&D programme need to be understood and dimensioned to the hazard posed by the waste inventory, following a graded approach.

The basic considerations to be taken into account when deriving WAC within the safety case include:

— Identification of requirements and constraints;
— Identification of disposal system components and their functions;
— Identification of key administrative and technical parameters associated with waste forms and waste packages that are relevant to the facility safety assessment and safety case;
— Quantification of acceptable limits or ranges for the technical parameters, in direct conjunction with the disposal facility design, site characteristics or licence conditions;
— Identification of acceptable and proven methods for calculation/measurement and verification of parameter values (and understanding of uncertainties involved);
— Identification of procedures for dealing with departures or non-conformities to the WAC;
— Documenting in a clear and concise manner the requirements and procedures and obtaining any necessary regulatory approvals for WAC;
— Distribution of WAC for review and critical comment, revising as necessary to respond to stakeholder feedback;
— Distribution of WAC and their implementation;
— Monitoring the effectiveness and suitability of WAC throughout the period when waste is accepted;
— Periodic revisions to reflect changing conditions and knowledge.

4.2.1. Identification of system components and their functions

Relevant information regarding waste types in the inventory for disposal, site conditions, disposal facility design and potential waste disposal arrangements, etc., need to be identified in order to provide input into the process for the development of suitable WAC. The impact on dose/risk of disposal system barriers and components operating in alone or in combination (waste form, waste package, engineered barriers, a natural barrier system and the geosphere) will have to be considered when developing the safety assessment that will serve as a basis for development of WAC. At an overview level, the disposal concept for low and intermediate level radioactive waste is based on the use of a multibarrier concept comprising the waste package, the engineered structures of the disposal facility and the nature of the host site, as well as institutional controls while they are operating. These aspects work in combination to achieve safety during disposal facility operation and closure, post-closure monitoring and institutional control (when present) and post-closure when passive safety alone is relied on. While safety during the first two phases is based on the nature of the waste package and possibly the performance of other engineered barriers, safety during the third phase relies also on the site characteristics. This multibarrier concept provides flexibility in determining the requirements for waste package acceptance, taking into account the fact that safety can be enhanced through the use of more durable and robust packaging and optimizing the disposal facility design and the choice of its location. Defining the requirements of a waste package by balancing these various features to achieve the disposal safety goals in all three phases of a disposal facility will permit an appropriate specification regarding acceptable radiological, physical, chemical and biological attributes associated with the waste including activity content, waste form, package configuration, etc., as described below.

4.2.2. Identification of key parameters

There are many considerations that need to be taken into account when developing WAC. Relevant considerations would include:

— The context (e.g. WAC for acceptance of raw waste for treatment, storage of waste packages, final disposal of waste packages, for an existing facility, for a planned facility);
— Legal and administrative requirements (e.g. waste classification and/or categorization scheme [2], reporting requirements, dose limits or constraints);

— The historical, current and anticipated waste inventory (location, waste streams, volume, nature, radiological, physical and chemical characteristics, production rates, etc.);
— The design of the waste management facility (description of functions, components, mode of operations, emplacement geometry, design limits);
— The disposal facility site and its characteristics relevant for safety (permeability, transport mechanisms and pathways, retardation potential, etc.);
— Operational procedures;
— The safety case and licensing basis of the waste management facility.

A list of properties that need to be included in WAC will be determined based on these considerations and their use in safety assessment calculations, which will then make it possible to define the criteria (qualitative and quantitative).

Note that one of the most important parameters is related to the radionuclide content, which is assessed based on the results of the safety assessment. Further details on how relevant content limits for near surface repositories could be derived can be found in a dedicated IAEA publication [23].

4.2.3. Quantification of acceptable limits

In detail, design and safety related requirements are facility-specific and no two disposal systems are exactly alike. It is vital to mention that copying conditions or parameter values from the WAC of other facilities (even ones that are superficially very similar), rather than developing case-specific criteria, can lead to the creation of inappropriate criteria. Therefore, the parameters to be used in WAC and their associated values and limits can only be determined in conjunction with the design and safety assessment for a specific facility. It is essential that all values and limits employed in WAC are based on the specific conditions of the intended facility.

WAC parameter values or limits can vary widely depending on the design and the role assigned to each of the safety barriers and how these barriers are implemented (as well as their behaviour in time) as discussed previously. For example, if the primary role of the waste container is just to act as a receptacle for holding the waste during handling and transportation, then limits on container durability, structural integrity, etc., could be very simple. Conversely, if the waste package is expected to provide structural support to prevent subsidence of the roof of a disposal facility cell, then it may require much more rigorous control of mechanical properties such as void fraction and compressive strength. Examples of the management of alternative packaging strategies are provided in Appendix I.

The facility type (e.g. processing, storage or disposal) also plays a significant role in the specification of relevant quantified values and limits. For example, WAC for a predisposal treatment or storage facility will typically include limits on surface contamination and activity levels associated with radionuclides that are important for worker health and safety (e.g. gamma-emitting nuclides such as ^{60}Co and mobile nuclides such as tritium). On the other hand, WAC for a disposal facility (repository) will need to take into greater account the impact of long-lived alpha and beta-emitting radionuclides that are important for long term safety and considered in the safety case. Reference [23] provides valuable information on how to derive the activity limits in the case of radioactive waste disposal in near surface facilities. The practice in France of deriving activity limits from the safety case is illustrated in Fig. 7. Limits may be calculated for the global capacity of the disposal facility, for each individual vault and for each waste package in the vault.

The maximum acceptable activity limits, which are determined based on safety assessment, are a major issue for the WAC. For example, these limits were defined for the Centre de l'Aube by two approaches (see Fig. 8):

— Operational safety and intrusion scenario for activity limit per package considering air transfer;
— Long term safety considering water transfer for global capacities.

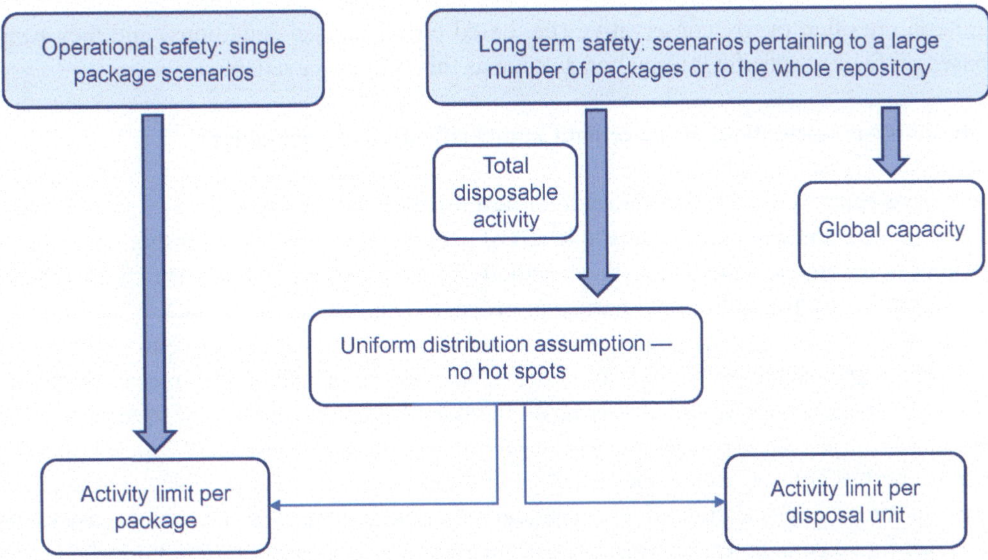

FIG. 7. Derivation of waste acceptance criteria for disposal in France. Adapted from Ref. [24].

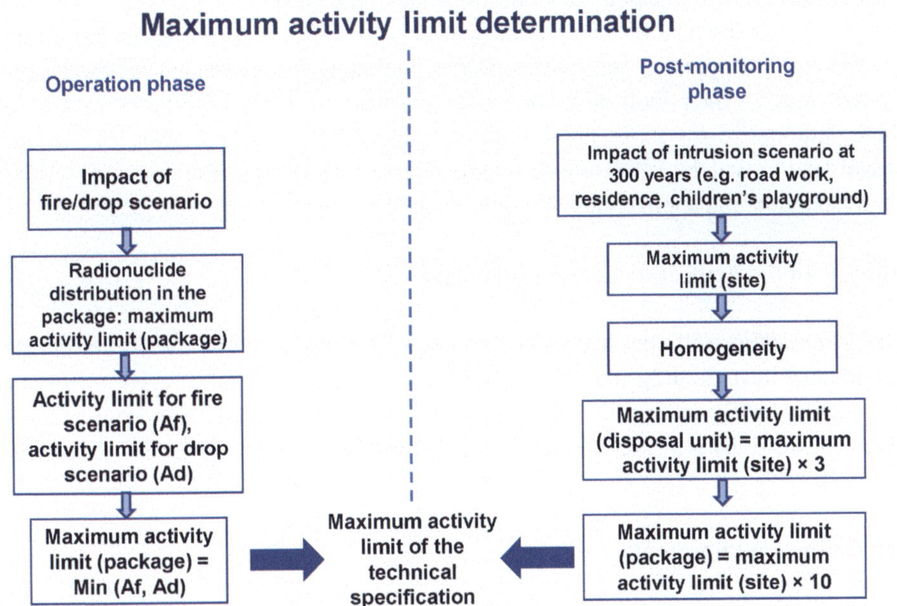

FIG. 8. Approach for determining the maximum activity limits based on waste acceptance criteria (radiological), National Radioactive Waste Management Agency (Andra), France. Adapted from Ref. [24].

The maximum activity limit determination for waste packages is derived from the global capacities (volume and activity). Homogeneity is considered for the whole disposal facility, with no hot spots allowed. This is achieved with activity limits for waste packages and for disposal vaults. Further details of the process for WAC development in France are given in Annex II.

Even in the absence of a facility design, generic requirements for different facilities can still be determined (e.g. for near surface disposal, sub-surface disposal, waste storage facilities and waste treatment facilities) and hence generic WAC can be developed (see Section 4.1 and Table 5 in Section 4.3.2.2), along with preliminary quantitative criteria associated with limits (Annex VII). However, such generic criteria are often overly conservative (i.e. based on worst case conditions) and they need to be revised based on facility-specific information as soon as this becomes available.

4.2.4. Methods for calculation, measurement and verification of parameters

When developing WAC, it is necessary to ensure that defined parameters and waste characteristics are amenable to measurement at the requisite levels of accuracy and precision. Furthermore, the users of WAC will have to develop and implement their methods for the measurement or calculation of the relevant parameters approved by the appropriate parties in order to demonstrate compliance. Consequently, it is required that WAC not include any conditions or limits for which it is not possible to show compliance. The specification of unachievable conditions in the absence of calculation or measurement methods suitable for routine application would clearly result in undesired complexities (e.g. significant delays and cost increases).

Some properties are easy to measure using assays and equipment that provide clear results such as chemical composition, physical dimensions, mass and radiation dose rates. Some are less easy to measure, such as residual liquid content, labile chemical content and non-gamma-emitting radionuclide content. In these cases, various calculation methods exist to infer the values from other measurements (e.g. use of scaling factors for difficult-to-measure radionuclides, sampling and measurement of raw waste streams along with process knowledge to calculate values in waste packages).

The WAC and/or associated documentation will give guidance and recommendations on acceptable methods for showing compliance with the criteria, including measurement methods or calculation approaches, procedures for their application, plus the identification, analysis and treatment of uncertainties. This is of high importance for those properties that cannot be measured easily or, depending on the assumptions, can lead to widely different results. Procedures that are specified in WAC will also have the extent of their availability and applicability clearly indicated.

4.2.5. Dealing with departures or non-conformities

Procedures for dealing with departures and non-conformities need to exist. Requirement 12 of GSR Part 5 [8] establishes in para. 4.26 that "The operators' procedures for the reception of waste have to contain provisions for safely managing waste that fails to meet the acceptance criteria; for example, by taking remedial actions or by returning the waste". This topic is discussed in detail in Sections 5.3 and 5.6 and in Appendix II.

4.2.6. Documenting requirements

The final stage of developing the WAC involves documenting the requirements in a suitable form for distribution to regulatory authorities, waste generators and upstream facility operators, as well as other interested stakeholders. This topic is dealt with in detail in Section 4.3.

4.3. DOCUMENTATION

Since there are many potential types of users (e.g. waste generators, facility operators, regulators, third-party inspectors and technical support organizations), the WAC documentation needs to be coherent, intelligible, unambiguous and usable by all involved parties. In the process of developing WAC, efforts towards achieving clarity need to be made. These will benefit the overall waste management process and reduce the likelihood of non-conformities. It is advisable to provide a clear explanation and justification of the respective criteria (including their basis and assumptions used) for the benefit of current and future stakeholders.

The WAC documentation, and related workflow processes, will have to be subjected to QA reviews on a regular basis.

4.3.1. Basic structure

The wide variety of requirements and associated criteria that are important in a WAC document can be grouped in a logical manner, such as:

(a) Administrative and quality control requirements:
    (i)   Application, agreement and acceptance process;
    (ii) Acceptable waste types, classes and forms;
    (iii) Indication of standard container types and waste forms, labelling/identification, tracking, etc.;
    (iv) Documentation requirements;
    (v)   Qualification process and statements of compliance;
    (vi) Quality management and QA certification requirements.
(b) Technical topics:
    (i)   Contents of waste package, e.g. limits on radiological content, criticality safety, chemical stability, fire safety, limits on toxic or organic materials, presence of free liquids, loose powders, gas generation rate limits and waste form to limit nuclide leach rates. WAC may include prohibitions, for instance pyrophoric, putrescible, burnable, chemotoxic and chelating reagents.
    (ii) Waste container design, e.g. container configuration, dimensions, waste package mechanical properties and package standardization.
    (iii) Operational (handling) matters, e.g. waste package condition, weight, handling arrangements and stacking.

These categories provide the initial structure for a WAC document. There are a number of ways the documentation can be organized and refined. In relation to the waste management strategy (inventory, type of waste, type of configurations, etc.) a single document could be used to contain all of the conditions, or alternatively, a set of documents might be needed to cover different aspects of the criteria (e.g. common administrative conditions in one document and technical/safety related matters, organized e.g. by facility or waste type, in separate documents appended to the main document).

The following subsections more fully describe relevant overarching issues to be addressed in the WAC documents, and these in turn can be used to provide a logical, more detailed content and structure.

4.3.2. Basic content of waste acceptance criteria

4.3.2.1. Administrative and quality control requirements

Administrative requirements are generally related to the application and agreement procedures and the tracking and reporting of wastes. They also include the QA requirements. These arrangements could include specifications for the labelling of waste packages or package identification and use, specification of a QA standard, definition of responsibilities, specification on duration of record keeping, and specification of format and content for documentation to accompany waste packages. These requirements are usually generic and can be specified without reference to a particular facility design or safety case. Demonstration that the required legal permits for the transfer or disposal of the waste are in place may comprise an essential component of the administrative requirements.

Quality related requirements are generally related to the application of QMSs, for example ISO 9001 [21], together with related procedures and certifications that describe QA of the waste and waste packages [5]. Appendix I provides information on the choice of an immobilization and packaging approach to meet package stability criteria.

Specification of the documentation requirements is an important part of administrative and quality control requirements in the WAC, since by this means, all relevant information is transferred to the waste receiver to permit the safe and effective future management of the waste. As discussed in Section 3.4 and Tables 2 and 4, the information to be passed from the waste generator to the waste receiver includes:

— Quantity of wastes and rate of arisings;
— Results of package monitoring, surface contamination, dose rate measurement and the results of other analytical techniques applied, including uncertainties and QA information;
— Container form and characteristics, including dimensions, material of fabrication, fabrication methodology, QA certification, corrosion and strength testing, method of closure, lifting attachments, etc.;
— Conditioning process and matrix characteristics, including certification of satisfactory process operation;
— Waste package content characteristics, waste form, weight, nuclide inventory, activity content;
— Waste package marking and index numbering;
— Waste package storage history;
— Demonstration of compliance with transport regulations;
— Independent inspection records (if required);
— QA certification;
— Disposal permits (if required).

A disposal site or a storage facility will typically be designed, and its safety case developed based on the use of specific types and sizes of waste container, together with associated specified waste forms which themselves will be based on radioactive waste inventory and its safety implications. The relationship between the waste form and the selected type of container influences the waste package performance (Fig. 9). Both the waste form and the container forming the waste package can be considered a dual engineered barrier as the waste container acts for the most part as a physical barrier (noting that corrosion products, for example, may be a beneficial influence on local redox conditions) as well as accommodating

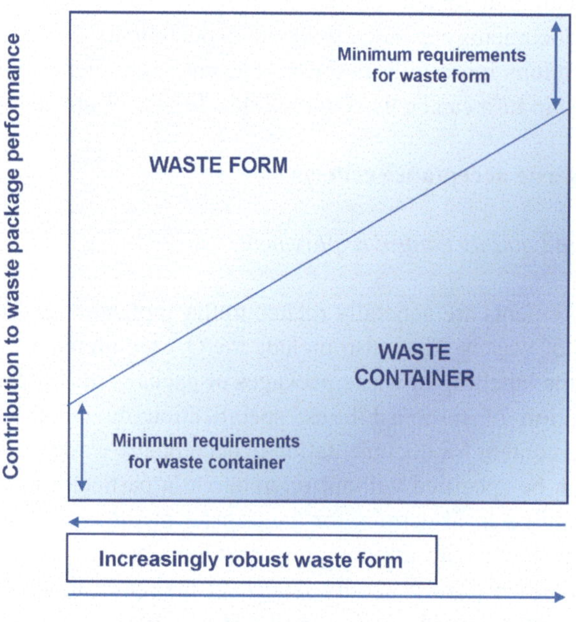

FIG. 9. *The relationship between waste form and waste container in terms of their contribution to waste package performance.*

safe handling, while the waste form more usually may provide both physical and chemical containment of the radionuclides in the waste.

The efficiency of disposal facility operations will be facilitated using standard waste package configurations and, where employed, these standardized waste packages are to be referenced as a quality control requirement in the WAC. Typically, for a relatively extensive inventory of LLW and ILW, standardized waste packages might comprise a mix of 200 L and 500 L drums, ISO storage containers, concrete or metal boxes with specific handling and stacking features, etc.

While it is good practice to use waste packages with standardized configurations, it is also good practice to allow flexibility by accepting non-standardized waste packages on a case-by-case basis, after due consideration of their technical and safety acceptability. This may be necessary to accommodate specific low volume waste streams or unusual circumstances.

4.3.2.2. *Technical aspects*

It is essential to understand the fundamental purpose of each technical requirement specified by the WAC. Safety related requirements are derived from the safety assessment calculations, normally for a specific facility. These may include limits on parameter values for particular radionuclides, requirements for conditioning by a particular approach, exclusions or restrictions of certain wastes or materials (e.g. liquids, explosives), etc. Design and operation related requirements are derived from the safety case and design basis for a particular facility. These include limits on waste package types, dimensions, mass, compressive strength, void space, etc. Since the design aspects feature in a safety assessment, they are closely linked and interconnected. Consequently, they can either be combined into a single group or split into two groups, depending on a specific case basis dictated by needs or preferences.

The design and safety related requirements that frame WAC need to be supported by a technical justification as to why they are important and why the particular limits or range of values for a given requirement were chosen. This aids the WAC users to better understand the intent of the requirement and how it can best be met. It also provides evidence to stakeholders that the WAC were developed in a logical and systematic manner, and that the requirements and criteria are clearly linked back to specific aspects of the facility design and safety case, as discussed earlier.

As an example (shown in Table 5), the technical requirements to be satisfied by waste package conformance with WAC can be grouped into the following five categories: radioactive content, stability, containment, radiation protection measures and traceability (record keeping). This structure facilitates

TABLE 5. THE FIVE MAIN CATEGORIES OF TECHNICAL REQUIREMENTS APPLICABLE TO WASTE ACCEPTANCE CRITERIA

Category	Description
Radioactive content	Requirements for the determination of the specific and total activity of the waste form and the waste package.
Stability	Requirements governing the physical, chemical and radiological characteristics of the waste form for the avoidance of (e.g.) fire, chemical reaction, criticality and mechanical collapse during transport, handling and stacking.
Containment: Measures for preventing or controlling the leakage and migration of nuclides	Requirements for containment of the radionuclides in the waste by the provision of a suitable waste form and container in order to prevent nuclide release from the waste packages during handling and transport and prevent/control release into the surrounding environment following disposal.
Radiation protection	Requirements for shielding, isolation, surface contamination, etc. for the protection of personnel from radiation exposure during handling and transport of the waste packages.
Traceability (records)	Requirements for record keeping so that the properties, history and quality management of each waste package can be traced.

the ordered development of the WAC. See Annex V for an example of the use of this approach for the determination of acceptance criteria in Japan. It is noted that the record keeping aspect overlaps with administrative requirements but is included here under technical requirements since it involves the generation and compilation of technical information.

It is important to note that not all of the requirements and criteria regarding waste forms necessarily apply to the WAC for all types of facilities. Typical requirements that are specified for various types of waste and facilities are summarized in Table 6. Appendix III provides some examples of WAC for various predisposal facilities and activities.

4.3.3. Non-standard and problematic waste

With regard to problematic waste, such as waste containing toxic or other hazardous materials, or non-routine wastes requiring extra consideration in treatment/conditioning, information on the physical, chemical and biological properties of the raw waste is necessary to be provided to ensure safety in waste processing, transport and acceptance of waste packages for storage and disposal. The quantity of non-radiological hazardous waste constituents generated in a facility will have to be minimized and their concentration determined and declared. Special provisions for keeping of records on chemo-toxic and other hazardous waste need to be made [24].

Attention is required for identifying, quantifying and documenting the presence of any of the following in a waste package:

— Sealed spent radiation sources;
— Free liquids and loose powders;
— Immobilized liquids with low flash point;
— Combustible materials;
— Nitrated ion exchange material;
— Explosive material;
— Toxic material;
— Corrosive material;
— Complexing agents;
— Salts and their concentrations;
— Strong oxidizing agents;
— Pyrophoric materials;
— Material causing chemical reactions including the generation of gases;
— Biological, pathogenic and infectious material.

4.3.4. Dealing with non-conformities

On occasions, waste packages may be received that, on inspection at the waste receiver's facility, do not comply with the WAC. Such cases are dealt with as non-conformities as opposed to by a departures request that is made by the waste generator in advance of waste dispatch. A departure relates to wastes with characteristics which will fall outside of the normal waste receiver's acceptance criteria and for which a special case needs to be made.

The waste receiver needs to anticipate the possibility of non-conformities by indicating potential approaches and procedures for resolution in the receiver's WAC documentation. These could include acceptance as a special case (following further assessment of impact on safety and operations), penalty charges and return to waste generator for rework. Further discussion on dealing with non-conformities and departure requests is provided in Sections 5.6 and 5.3, respectively, and in Appendix II.

TABLE 6. SUMMARY OF TYPICAL WASTE ACCEPTANCE REQUIREMENTS AND THE INTERFACE WITH DIFFERENT TYPES OF FACILITIES

Requirement type	General requirement	Waste generator	Waste processor	Waste storage operator	Transporter	Disposal operator
Administrative	Waste/container identification and tracking	✓	✓	✓	✓	✓
	Use of standardized containers		✓	✓	✓	✓
	Reporting requirements	✓	✓	✓	✓	✓
Quality/qualification	Characterization requirements	✓	✓	✓	✓	✓
	Inspection and test requirements on incoming waste packages			✓	✓	✓
	Acceptable waste types/classes/forms		✓	✓	✓	✓
	Restrictions on or specification of allowable conditioning methods		✓	✓		✓
Technical: Design/operational related	Minimum/maximum dimensions		✓	✓	✓	✓
	Maximum mass		✓	✓	✓	✓
	Mechanical properties of waste form/package			✓	✓	✓
	Waste package handling		✓	✓	✓	✓
Technical: Safety related	Waste form chemical stability			✓		✓
	Waste form radiation stability			✓		✓
	Waste package/container durability			✓	✓	✓
	Restrictions on waste form			✓		✓
	Restrictions on chemical or/and hazardous constituent content	✓	✓	✓		✓
	Restrictions on biological, pathogenic and/or infectious material content	✓	✓	✓		✓
	Restrictions on free liquid content			✓	✓	✓
	Restrictions on combustible material content	✓	✓	✓	✓	✓
	Restrictions on heat generation rate	✓	✓	✓	✓	✓
	Restrictions on radionuclide content	✓	✓	✓	✓	✓
	Restrictions on fissile content	✓	✓	✓	✓	✓
	Gamma radiation dose rates acceptable/authorized levels	✓	✓	✓	✓	✓
	Restrictions on fixed and/or removable surface contamination on waste containers/packages	✓	✓	✓	✓	✓

4.3.5. Acceptance of historical waste

Departures are often sought for wastes that were produced before the existence of the WAC or while previous WAC with different criteria were in force (e.g. legacy or historical wastes) or for small quantities of wastes, for which all the information required for waste acceptance at a downstream facility cannot be provided.

Acceptance of these waste packages for long term management will need to follow a case-by-case approach, as they cannot be judged on pre-established requirements such as WAC. The assessment process will rely on all available information, the number of waste packages concerned and their hazard. It is advisable to perform the waste characterization as far as possible. If the waste is similar to other well characterized waste, this could be taken into account. If, nevertheless, the waste is still non-compliant, and therefore not disposable, additional treatment, characterization, conditioning, etc. will probably be needed.

Problematic characteristics of historical waste include:

— Poor or no historical data and traceability;
— Unclear waste origin;
— That they may be conditioned, partially embedded, or bulk;
— Degraded condition;
— Mixed waste streams.

4.4. FURTHER CONSIDERATIONS

4.4.1. Considerations on developing waste acceptance criteria based on the availability of a disposal facility

If a disposal option is not available, it is advised that the IAEA waste management principles be considered while developing WAC. A prerequisite is to have detailed characterization and documentation of the waste as well as a developed safety case prior to the development of WAC. It is to be noted that requirements need to be put in place to ensure that a waste package will not be a source of problems during the expected duration of the storage phase, taking into account the ageing of materials (drums, lifting equipment, matrix, etc.) and mechanical properties (weight, resistance). Even though waste is generated under existing regulatory and environmental criteria, there is a risk of a change in these requirements that may lead to a need to re-characterize and/or further condition stored waste.

Where no other option is available (e.g. clearance, recycling or export), the final destination of radioactive waste is disposal in a suitable disposal facility. The waste needs to be in a form that is compatible or can be made compatible with disposal facility WAC. Two cases have to be considered (see Fig. 10). In the first case, WAC for the disposal facility exist or at least have been drafted on the basis that a disposal facility exists while in the second case, those criteria have not yet been defined.

In the first case, conditioning of the waste packages can be based on the requirements derived from both storage and disposal. Conditioning steps can be qualified for both of these stages, and the subsequent transfer of waste to a disposal facility would not require extra conditioning or characterization steps to be made.

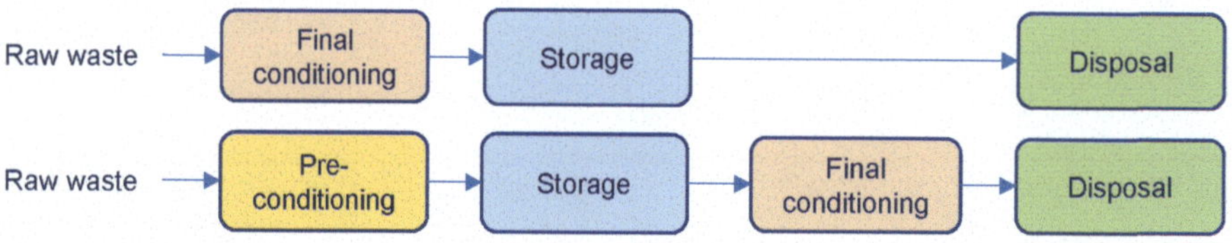

FIG. 10. The relationship between storage and the disposal and conditioning steps.

In the second case, where disposal facility requirements are not yet well defined, the conditioning of the waste packages cannot rely on any existing specific requirements for disposal. As most raw wastes cannot generally be stored safely for long periods, preconditioning steps may be needed to produce a product with the necessary stability and other characteristics to allow storage.

In the design of these preconditioning processes, care will be paid to avoid steps that could later lead to difficulties when disposal facility conditions are established (e.g. incompatible package design, re-conditioning needs, waste forms that cannot be brought in line with disposal facility conditions). However, ensuring such flexibility during preconditioning steps will have to be carefully balanced not to compromise the safety issues during storage and not to lead to extra costs.

The first case undoubtedly has safety, technical and economic advantages since it reduces uncertainties for waste owners and disposal facility planners and promotes a more straightforward integrated waste management scheme. Therefore, basic planning for the complete life cycle of the waste, from its generation to its disposal, and defining preliminary criteria for later disposal is helpful, even though no guarantees can be granted. As discussed in Section 4.1, the development of WAC is an iterative process; when downstream information is not available such as that pertaining to disposal requirements, then WAC would have to anticipate these requirements as much as possible. However, as and when that information becomes available, the WAC can be refined as appropriate.

4.4.2. Uncertainties and conservatism

WAC need to encompass all requirements necessary for the safe operation of a waste management facility and for the post-closure phase of a disposal facility. In their preliminary form, requirements may, as earlier noted, *qualitatively* specify the need for measures to be taken to achieve the safety objectives of waste management and define the principles by which these objectives will be demonstrated. WAC reflect the need for such measures.

The resulting criteria will have to be established in such a way that, on the one hand, they are precise enough to ensure in their entirety the required safety of a facility and its operations, while on the other hand, they might be fulfilled in a more flexible way if required. Understanding requirements and translating them into WAC means taking into account the purpose and role of each of the barriers in a disposal system without overly using them beyond what is required in meeting the objectives (e.g. a high-quality waste form in a high-quality container in a high-quality facility on a high-quality site). Such an approach offers flexibility to the waste generator or operator of the facility in the selection of appropriate waste processing and packaging methods without affecting the required safety of the storage facility or the disposal facility. Examples of appropriate waste processing and packaging methods are given in Appendix I.

There will always be a number of uncertainties inherent in setting criteria and/or the measurements taken to show compliance with the criteria. The specification of quantitative limits, at least applied to the activity limits, will take into account the possible uncertainties and their impact.

Criteria that are derived from design or safety assessment information will depend on the quality of the original data and analytical assumptions, on the computer codes used for analysis and on the sensitivity of assessment results to particular input parameters. In cases where the result (e.g. dose or risk) is very sensitive to the input data and assumptions, a conservative specification for a waste acceptance criterion is often used to ensure that the safety or design property is not exceeded. In cases where the result is not very sensitive to the input data and assumptions, more uncertainty can be tolerated, and less conservatism needs to be applied when defining WAC.

Another source of uncertainty arises from application of the WAC based on results from measurement or calculation techniques applied to the waste packages. The application of different techniques will have a range of inherent uncertainties associated with them, in terms of operator failings, measurement accuracy and precision, representativity, etc. It is important to note that any parameter specified in the WAC for measurement needs to be justified by facility or process requirements and be capable of being reasonably quantified (measured or calculated) to the limit of requested accuracy (as earlier discussed in Section 4.2.3).

For example, if a particular radionuclide limit is specified but cannot be measured or calculated to that level or required degree of accuracy, then it will not be possible to show compliance with the WAC. In such a case, excessive conservatism may be applied but this may significantly reduce the amount of waste that can be disposed of at a particular disposal facility. This is because the inventory of some nuclides could reach the capacity of the disposal facility much sooner than if actual values were used. Requirements for unjustified parameter accuracy and precision can also result in an improper or ineffective application of WAC.

To summarize, conservatism in the use of parameters and the approaches used to determine the characteristics of waste to be disposed of depends on the uncertainties inherent in:

— Data relating to waste characterization;
— Data relating to the site characteristics;
— Data relating to the engineered system, in particular the barriers;
— The method used to obtain and analyse data;
— The approaches and models used to determine uncertainties.

It is important to set reasonable WAC that are possible to verify. Numerical limits would be defined for the significant parameters and unnecessary conservatism needs to be avoided.

4.4.3. Waste acceptance criteria review

Once the WAC have been formulated and approved, periodic reviews are required in order to identify issues and opportunities for improvement and as a result update the WAC. Such periodic reviews would incorporate knowledge concerning any new developments that may affect the criteria, such as:

— Changes to predisposal processes;
— Amendment of disposal WAC;
— Evolution of regulatory constraints/licence conditions;
— Production of new waste streams that are not fully addressed by existing criteria;
— Changes and updates of the safety case and results of periodical safety reviews.

When WAC are revised, the resultant consequences need to be carefully evaluated. For instance, if new criteria are introduced, it is possible that wastes already accepted at a facility (e.g. in a store or disposal facility or for a treatment facility) would not conform with the revised WAC. The facility would need to identify alternative conditioning/treatment options. Application of new requirements to wastes that have not yet been disposed of is generally feasible but often results in a cost penalty, increase of waste volume and additional occupational dose (from re-conditioning, re-packaging, etc.). Discussions and negotiations will have to take place between the waste generator and processors to ensure the optimum waste management solution. For instance, it may be more efficient to adapt the facility design to address the consequences of new constraints rather than impose new or tighter acceptance criteria.

In most cases, the new WAC conditions are not applicable to already-disposed-of waste in closed disposal units, since disposal is supposed to be an irreversible step with no intention of retrieval after closure. However, if it is realized that the previously conducted safety assessment did not consider certain information and that the WAC in use are not fully appropriate, for example, if some safety-critical nuclides were not addressed or gas generation was overlooked, then the operator will have to revisit the information on the already disposed waste and evaluate the safety consequences through a revised safety assessment. This may result in changing the WAC for future waste and a decision on re-working may be needed for the disposed waste (or a proportion of it).

WAC revisions need to be undertaken in a controlled manner, with the changes from previous revisions clearly identified and explained. This helps the user easily understand what changes have been made and why. It is important that the revised WAC are issued to all users in a timely manner to ensure that they all have the same information.

5. IMPLEMENTING THE WASTE PACKAGE ACCEPTANCE PROCESS

A waste package is often utilized by design as one of the engineered barriers for providing one or more safety functions, not least containment. It also represents a principal unit of conditioned waste for administrative and practical purposes and is used as a reference for recording and controlling information and for making decisions.

The acceptance of any waste package requires compliance of this waste package with both the 'agreement' (qualification) and the on-receipt inspection, monitoring and control processes in order to maintain confidence. The process of acceptance is illustrated in Fig. 11.

It is preferable that WAC exist before waste is generated or processed and before waste is transferred from one facility or institution to another. Similarly, the qualification process needs to be completed before the submittal of a request for acceptance of waste packages. Additionally, any departure request will have to be resolved before the waste is produced (see Section 5.3 for further discussion of departures).

During the process of on-receipt compliance checking, the treatment of non-conformities needs to be addressed at an early stage, since decisions and corrective measures can often be taken in order to treat the non-conforming waste. Section 5.6 discusses the handling of non-conformities in greater detail. At the end of this process, the acceptance (compliance verification) of waste packages can be granted.

The process of agreement for acceptance between entities could be a mere formality if:

— Waste packages have been conditioned according to a qualified acceptable procedure.
— There is demonstration of conformance of waste packages to specification of WAC.
— There is traceability from generator documentation to final individual waste package.

5.1. AGREEMENT

Agreement comprises an important aspect of the administrative process, dealing with an application by a waste generator for the acceptance of wastes or a waste stream at a receiving facility, whether for treatment, conditioning, storage or disposal. If the waste transfer is taking place simply between two operating units on the same operating site, or within the same organization, the agreement process can be simple and may not need formal WAC, instead consisting of necessary information requirements and meeting internal management procedures. However, between two different organizations, the agreement process will need to be more formalized and needs to be stipulated in the WAC documentation suite. It is likely to comprise a contractual arrangement.

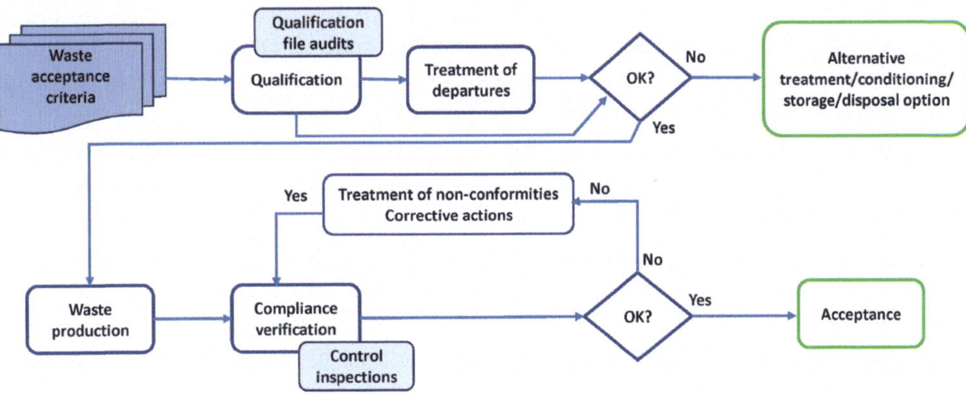

FIG. 11. The waste package acceptance process.

The application/acceptance process consists of several steps:

— *Application by the waste generator for transfer of wastes* – The application will be in a format specified by the waste receiver to ensure it includes all the information pertaining to the waste that is required at this stage. This may include traceability documentation from waste generation to a final individual waste package, which includes, for instance: description, radionuclide inventory, details of radiological content, dose rates, presence of hazardous materials, immobilization matrix, QA reports and certification, including analytical reports and statement of uncertainties, and necessary permitting for treatment/disposal from the regulators and other documentation as required.

— *Agreement in principle* – The waste receiver may provide, on inspection of the application, agreement in principle to accept the wastes, which may be provided with qualifications, for example, on the provision of more information to clarify specific aspects of the wastes. An agreement in principle may be sought by the applicant when the wastes to be committed are not yet in a form ready to be transferred or when waste is anticipated. The latter may arise when a new facility is being planned or constructed and assurance is sought that the anticipated wastes can be accepted by the downstream facility, with guidance on acceptable waste forms, containers, lifting arrangement, etc., being particularly helpful in finalizing pre-processing arrangements. The agreement process could lead to a formal agreement if the qualification of the waste package confirms compliance with the acceptance requirements and relevant QA procedures.

— *Formal agreement* to receive the wastes on shipment to the waste receiver's facility. This may follow a further application for acceptance of waste for which agreement in principle was received, but now is in existence and compliant with the acceptance criteria. This agreement may take the form of a commercial contract, formalizing handover documentation requirements and providing the transfer of title to the wastes (e.g. for disposal).

— *Acceptance receipt* – On delivery of the wastes to the receiving facility, it may be appropriate to provide a simple confirmation of delivery or, after suitable inspection and confirmation that all is in order, an acceptance receipt summarizing details of the wastes received.

Is important to highlight that while title to the wastes may have been transferred to the waste receiver, it is in the interest of the waste generator to maintain full records of wastes generated and transferred. This is both to meet stakeholder needs (e.g. regulatory audits) and in case there is a request from the downstream facility (e.g. store or disposal site) for more details on a given package (e.g. if an 'event' has occurred in respect of the package that needs further investigation).

5.2. QUALIFICATION

The agreement process leads to an operational definition of a waste package type that will be accepted, subject to it complying with the acceptance requirements and relevant QA procedures. The qualification approval expresses that the receiving waste management organization considers that waste packages produced by the agreed process will comply with the requirements (a priori acceptance). Qualification is a significant component of the agreement process. It provides assurance to the waste receiver that the relevant process, equipment or waste product can fulfil the specified requirements in compliance with the WAC.

The qualification of the waste management processes needs to demonstrate:

— That processes and equipment are operating in accordance with process design technical specifications and parameters;
— That important process parameters, variables and controls have not changed from those values established in the original process design (or if they have changed, they are not affecting WAC);
— That the related procedures that direct the waste processing, waste package production and storage were recorded, assessed and approved.

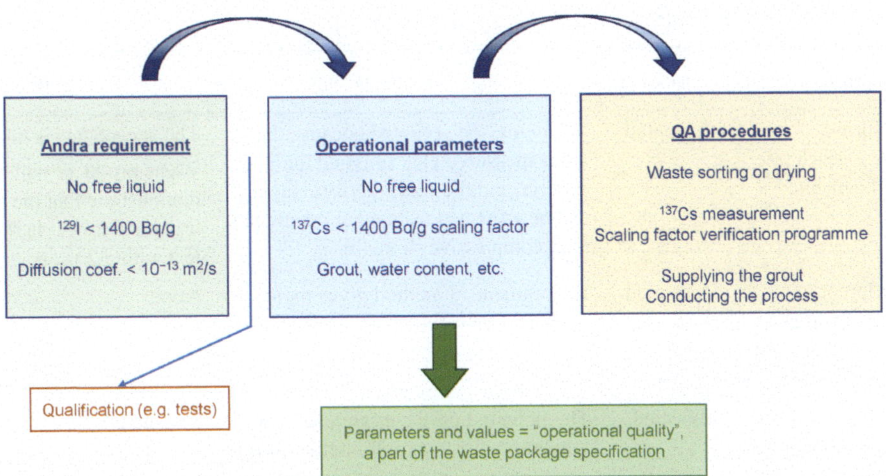

FIG. 12. *Illustration showing how French National Radioactive Waste Management Agency (Andra) requirements are translated into operational parameters, as well as the related quality assurance (QA) procedures. Adapted from Ref. [25].*

The following QA aspects are of particular importance for consideration:

— Inspections and controls, frequency of measurements and testing, methods and criteria;
— All steps required by the process verification (checklists, route cards, etc.) to confirm the completion of the critical process steps;
— Methods (incorporating an assessment of associated uncertainties) used to verify that the characteristics of incoming waste comply with the WAC.

Internal verification (through inspections, audits, etc.) are undertaken to demonstrate that conditions that could have a negative impact on the waste and waste packages' final quality are appropriately monitored and controlled and that corrective actions are implemented where required. When the processes are outsourced (e.g. manufacturing of waste containers) quality control arrangements need to be in place.

A compliance document that demonstrates adherence to the WAC needs to be established when all involved stakeholders have not yet agreed on compliance processes (i.e. as discussed in Section 5.1). The compliance document can either contain the list of all the criteria together with justification for their use or just indicate the qualification process agreed upon by the stakeholders.

As part of the qualification process, investigations are performed aiming to identify parameters (that can be monitored during the operational phases) associated with the conditioning process or with the package, to ensure and illustrate compliance with WAC. Figure 12 shows examples of how specific acceptance requirements are translated into operational parameters and the related QA procedures. For the diffusion coefficient example, these operational parameters may reflect the composition and the preparation of the concrete used to make the diffusion barrier, including its water content, type of cement, additives, etc. The data obtained during a qualification step have an impact on the choice of the operational parameters and QA procedures. In addition, experiments can be performed on prototype waste packages or samples. These investigations are performed by the waste generator or waste processor in dialogue with the waste management organization. Another example based on experiences in Belgium is discussed in Annex I.

5.3. DEPARTURES

In the exceptional case that a waste generator or waste processing operator is unable to comply with one (or more) WAC and this is recognized before the waste packages have been produced, the disposal operator may grant a departure, possibly under new conditions, if it can be justified and documented that

TABLE 7. EXAMPLES OF DEPARTURES REQUESTED

Reason for departures	Decision	Comments	Conditions
Inability to perform a technical test (e.g. leach rate measurement of cemented waste form)	Accepted	Waste may be accepted because the other measurements satisfied the criteria, and the waste performance can be estimated from other parameters (e.g. compressive strength).	The acceptance is time limited. Waste producer will have to demonstrate that the acceptance criteria are met via the necessary test within 2 years.
The use of a different technical characterization method from those that were recommended or required	Accepted	Recommended method gives more conservative results.	None
Disposal of pulverized waste (i.e. loose-powder-containing)	Rejected	The potential dose consequence via inhalation due to a malfunction during operation (e.g. in a drop scenario) does not meet the Andra protection objective.	Rejected

this departure is such that it has no negative influence on the further safe management of the primary waste package and would not impact on the safety of already received waste. (Note that a departure request would normally relate to a series of packages yet to be produced, rather than a once-off exception, which would be dealt with as a non-conformity.)

A procedure to deal with these departures will have to be established. An example of a procedure and of the minimum information requirements is given in Appendix II.

If the need for a departure is anticipated, before producing the waste packages, the waste generator or processor has to submit this request to the disposal operator (in some cases, the storage operator) to explore the possibility to not adhere to one part of the WAC. This could affect the production of a complete sequence or family of waste packages.

Setting realistic WAC that can be met in most cases while maintaining compliance with safety case requirements to minimize the occurrence of departure requests is of great importance for the receiving facility (e.g. stores or disposal sites).

It is advisable to treat departures on a case-by-case basis. The number of waste packages that will be approved for acceptance under departures has to be limited and they have to not have been produced prior to the acceptance of the departure. Table 7 illustrates some examples from the French National Radioactive Waste Management Agency (Andra) where requested departures were either accepted or refused.

5.4. QUALITY CONTROLS

It will be important to both the waste generator and waste receiver to establish a QA strategy and implementation plan for the waste, covering wastes being prepared for dispatch and for receipt. Procedures, instructions, checklists, route cards, etc., will be utilized by the operators in order to verify and produce objective evidence on the quality of the information associated with waste packages. Waste management facilities need to demonstrate that instrumentation used to monitor and control waste management operations are maintained and calibrated at predetermined intervals by accredited or QA-approved laboratories and by using reference calibration standards traceable to a recognized standard. Reports on material, chemical and radionuclide analysis will be issued by accredited or QA-approved laboratories. Inspection, measuring and test reports would be made available as part of the waste package data file, appropriately signed and dated with reference to the relevant instructions.

Important actions that need be implemented to verify compliance with the WAC are described in detail in Sections 5.4.1–5.4.3.

5.4.1. Internal and external audits

Audits are a key mechanism for establishing confidence in the quality of the waste packages delivered. Arrangements to be discussed between the various operators will include a combination of both internal and external audits. Internal audits are conducted by the organization responsible for a waste management facility based on its own systems and procedures, etc. External audits may be conducted by the entity receiving the waste or by a third party (e.g. a certified assessor ensuring compliance with external standards). The applicability of internal and external audits may vary from one Member State to another, but the overall objective is the same. The emphasis is on ensuring waste product quality and enhancing the confidence of all relevant stakeholders, including the public, to provide assurance that a waste package will perform as intended.

5.4.2. Inspections

Provision for verification through inspection and testing before waste packages are accepted for further management are to be established through WAC:

— It is advisable to perform reshipment inspections on waste packages at the waste generator or processor's premises before waste packages are shipped to the next stage. This can be a preventive measure allowing for early detection of non-conformities that can eliminate time consuming and costly rework. During pre-shipment inspections, particular attention needs to be paid to:
  - Verification of data files including checklists, reports of laboratory analysis, radiological data and measurements, etc.;
  - Visual damage to a waste package such as corrosion, cracks, markings, leaking, deformities and other possible mechanical damage.
— Receipt inspections and tests will be performed at the receiving facility on delivery of the waste packages. This aspect of WAC verification may include:
  - Completeness of documentation;
  - Quality checks (e.g. visual inspections);
  - Radiological checks for surface contamination and gamma ray emissions;
  - Waste package content inspections involving measurements;
  - Destructive and non-destructive testing and analysis;
  - Approved authorization.

A typical checklist for on-receipt checks of waste at a disposal facility is given in Table 8; similar checklists will appertain to other facilities such as stores or conditioning plants.

5.4.3. Destructive and non-destructive testing of waste packages

Waste packages can contain homogeneous or heterogeneous encapsulated waste. To guarantee that waste packages fulfil the specifications set out in the WAC, it is required that the waste form adhere to specified requirements such as:

— Mechanical strength;
— Durability;
— Leach resistance for homogeneous wastes or diffusion resistance for heterogeneous wastes;
— Physical integrity.

Waste characterization can be undertaken using different methodologies, as illustrated in Fig. 13.
Some parameters that are specified in WAC can be obtained during processing prior to packaging. Others may be measured during inspection at a facility, either directly on the surface of a waste package

TABLE 8. TYPICAL CHECKS AGAINST WASTE ACCEPTANCE CRITERIA ON WASTE RECEIPT AT DISPOSAL FACILITY

Administrative checks	Visual checks	Direct measurements
Completeness of consignment record	Package type	Weighing
Package identification	External package condition	Radiation dose survey
Weight	Tampered seals	Radiological contamination
Activity	Package closure	Closure tightness (torque) testing
Dose rate	Package labelling/identification	Radiography/tomography
Shipment number		(Gamma, neutron, X ray, etc.)
Surface contamination		Activity measurement
Special conditions		Container integrity survey
Container type		Destructive testing
Fissile mass		Compliance with transport regulations

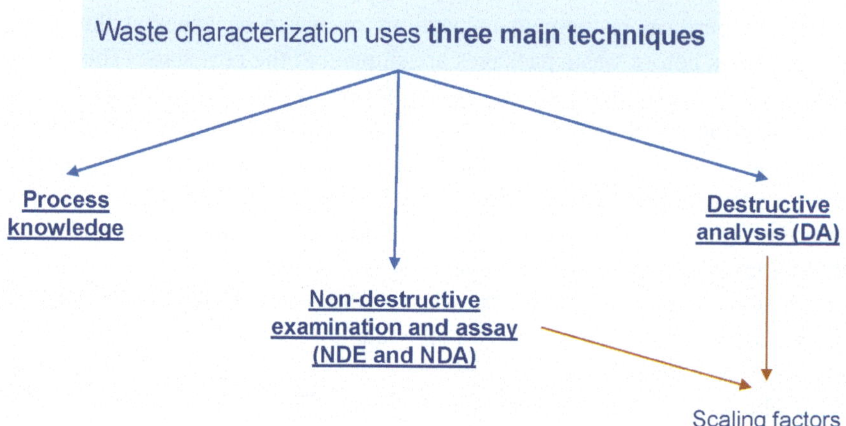

FIG. 13. Waste characterization methodologies.

or inside it using sensing and imaging technologies (non-destructive testing). For instance, the weight or dose rate of a package can be measured and compared with values reported in the data file.

There are some parameters, however, that cannot be evaluated directly during the processing of a waste package, for instance, waste form parameters such as leach rate, diffusion coefficient, etc. For these parameters it may be necessary to have direct access to the contents of a waste package in order to perform measurements and take samples for assay or experiments, some of which may last more than one year. This type of analysis, termed destructive testing, is considered a second level control measure based on a random sampling process.

Examples of destructive and non-destructive testing and control measures include:

— Non-destructive testing, which may consist of identifying the waste content with instruments such as X radiography, densitometry or conducting activity measurements with gamma spectrometry, passive and active neutron counting, emission computed tomography and photo-activation.
— Destructive testing, which may be performed on core samples to check the waste content or the quality of conditioning such as mechanical strength, diffusion coefficient and leach rate.

In the absence of a formal waste package testing programme (e.g. destructive and/or non-destructive testing), reliance is placed on process understanding and administrative control measures to confirm the quality of waste packages presented for disposal. Receipt inspections are conducted on the transport vehicle (e.g. visual inspection of the external waste package condition) and the accompanying documentation, and they are aimed at identifying potential non-conformities prior to off-loading and emplacement operations. Receipt inspections conducted at the disposal facility are usually a continuation of similar inspections performed by the predisposal operator in order to ensure that waste packages have not been damaged during the transport to the disposal facility. Further information on radioactive waste characterization and analytical methods, both destructive and non-destructive, may be found in Refs [17, 26].

5.5. VERIFICATION AND ACCEPTANCE

The on-receipt waste acceptance process includes verification of compliance with WAC after waste package inspection (see Section 5.4.2) and checking that the waste package documentation file is accurate and complete, as required by the WAC. The effective control of a documentation file supports confirmation of the quality of a waste package produced in a qualified conditioning facility by a qualified conditioning process and qualified methodologies for radiological, physical chemical and biological characterization.

5.6. NON-CONFORMITIES

In cases where a waste package is discovered to be non-compliant with the applicable WAC (e.g. on inspection prior to shipment or through testing and measurements on receipt), rectifying any non-conformity issues will have to be taken care of prior to any further management activities (as initially discussed in Section 4.3.2.4). For non-conforming individual waste packages there are generally three ways to proceed, for instance:

— Refuse acceptance;
— Acceptance after corrective measures have been undertaken successfully;
— Acceptance without taking corrective measures, on condition that proof is provided that the non-conformity will not jeopardize the safety of the facility and the further management of the conditioned radioactive waste.

The responsibility for providing this demonstration of proof that safety is not going to be jeopardized will ultimately rest with the waste generator or processor, but the waste receiver will have to confirm that this demonstration satisfies both their own criteria and those of the regulator, in anticipation of regulatory audits at a later stage. A documented procedure will be established in order to treat these non-conformities. Reference [2] provides helpful guidelines in this respect. An example of a procedure and of the minimum information to be provided to the disposal operator is given in Appendix II.

In cases where one or more packages do not meet the WAC, the number of packages would generally be limited, based on an assumption that a robust management system is in place. The waste generator or processor is requested to explain the root cause of any non-conformity (e.g. industrial malfunctioning, handling problem, external temporary storage) and submit a plan for approval before the waste is shipped onwards.

Non-conformities can be categorized according to their potential impact and severity, expressed in terms of a hazardous situation, liability, etc., and against the conformance domain that the non-conformity does not comply with. In general:

— Minor findings with no safety implications to the public, personnel or the environment (e.g. damaged or illegible placards, minor scratches to or dents in metal waste packages), can be brought to the attention of the predisposal operator by forwarding a copy of the completed inspection report.
— Significant non-conformities need to be notified to the predisposal operator so that appropriate preventive and corrective actions can be taken. Then, any affected waste packages are returned to the

originating operator as soon as is feasible or they are temporarily stored in an appropriate area until such time as the non-conformity can be addressed on-site, so that after remedial actions the issue can be closed out (e.g. as a result of repair, decay of dose rate or re-packaging).

Some non-conformities will require careful examination to better understand their implications, for example:

— Findings with real or potential safety implications to the public, personnel or the environment;
— Findings that may jeopardize the long term safety performance of the disposal facility (e.g. obvious damage resulting in leakage from containers);
— Non-compliance with respect to transport regulations identified while still on the predisposal site.

The measures taken in all cases of non-conformity need to be clearly communicated to the predisposal operator and in some cases agreed with the regulator. The corrective actions resulting from non-conformities need to be commensurate with the hazard and will address the following:

— Re-working a particular non-conforming waste package to a condition that is acceptable (could be performed by the predisposal operator or by the disposal operator for a fee);
— Reviewing and adjusting any processes that are responsible for generating the defective waste package to prevent the future generation of non-conforming packages due to the same cause;
— Carrying out remedial work to ensure conformity (e.g. removal of minor contamination spots on the external surface of the package, fixing minor dents on the package surface).

Informed decisions regarding the management of a non-conforming waste package can only be taken after advice from appropriate experts (scientists, engineers, safety assessment experts, etc.), discussions with the disposal operator concerning operational aspects (handling, risks, etc.), and post-closure performance (e.g. based on an assessment to evaluate dose or risk), and by taking into account lessons learned from the resolution of other similar (earlier) non-conformities. In some instances, regulatory authorities may need to be involved in discussions to make them aware of issues and to seek their guidance in finding an acceptable resolution. Some examples of non-conformities and remedial actions are illustrated in Table 9.

All investigations, causal analysis, corrective and preventive actions need to be suitably documented and will be added to the waste package information. Traceability of the actions taken to deal with non-conformities is also important and the recording of lessons learned is a key task within the objective of continuous quality and safety demonstration improvement, also with the aim of optimizing the WAC and minimizing the numbers of non-conformities. A non-conformity 'accepted under condition' will be followed through and the corrective actions documented since related information about the waste packages accepted under this process may be requested by involved parties.

TABLE 9. EXAMPLES OF NON-CONFORMITIES

Reason for non-conformity	Decision	Comments	Conditions
Damaged container concrete (e.g. a small indentation or chip due to a handling problem)	Minor non-conformity; accepted	Remaining thickness of the container still fulfils the containment functions/requirements	None
Inadequate mechanical strength of a metallic package	Significant non-conformity; initial rejection but could be accepted after rework	May impact the global mechanical stability and does not fulfil containment requirement	Reconditioned into a concrete container and re-evaluated for approval

A periodic review of the non-conformities will be carried out with the following objectives:

(a)   To estimate the frequency of occurrence of non-conformities according to their severity;
(b)   To identify trends and common causes;
(c)   To estimate the effect of preventive actions;
(d)   To consider whether it would be beneficial to redefine how non-conformities are classed;
(e)   To consider whether a revision to the acceptance criteria may be warranted.

The designer and, later, the operator of the facility, will take into consideration the risk of receipt of undetected non-conformities by the waste generator. As a precaution, since it may not be reasonable to exclude potential quality control problems with regard to waste packages, potential package defects remaining after disposal will be taken into consideration in scenarios used in the safety analysis as early as the design stage and especially in impact assessments.

6. CONCLUSIONS

It can be concluded that due to the integrated nature of the low and intermediate level RWM process, the interfaces between the different waste management entities (i.e. waste generator, waste processor, disposal operator and interfacing between operators and regulators) are of primary importance. WAC are quantitative or qualitative criteria for the waste form and waste package to be accepted by the operator of a waste management facility and they play a key role in managing interfaces between the entities safely and effectively by establishing clear responsibilities and controlling expectations. The WAC are also a valuable tool for ensuring the implementation of coordinated technological procedures throughout all stages of the radioactive waste life cycle.

It follows that the development of WAC needs to be performed in a structured and methodical manner, taking into account all relevant considerations including the regulatory context, design constraints and the need for safety during facility operation and the disposal facility post-closure period. In developing WAC, full consultation will be undertaken with key stakeholders, including waste generators, processing and storage facility operators and the regulatory authorities. Third parties, such as those with supporting R&D programmes, may be involved in the WAC development to confirm waste form and waste package performance. Where a country has not yet developed a disposal facility, generic requirements will be established for all steps in the waste management system. These will be detailed enough to allow for safe radioactive management practices, but not so prescriptive that they foreclose future management options. The final list of criteria resulting from recognized requirements will be developed on a case-by-case basis from first principles.

Development of WAC for low and intermediate level radioactive waste needs to follow a systematic approach taking into account the functions of the different components of the waste management system, the waste form, the waste container, other engineered barriers and natural barriers (e.g. host location for a disposal facility) and safety and operational constraints. It is appropriate to develop the WAC for disposal commencing from the performance based generic requirements and moving to the specification based requirements according to the progress of design stages. Failure to do so may result in WAC that are overly conservative and/or are not practical to enforce.

Since WAC are defined in relation to the properties of the waste, building knowledge of the inventory, in particular the characteristics of the waste, is a key task that need to be properly addressed. It is advisable to initiate characterization of waste as soon as possible, particularly for raw waste (i.e. before waste processing starts). Similarly, the development of a waste tracking system incorporating waste inventory data from the very beginning of waste generation is a useful tool to ensure inputs are provided in a timely manner using appropriate formats, terminology, etc.

In the context of deriving WAC, defining the function of the waste package is a key component of the safety assessment and safety case (e.g. whether it provides shielding in a store or long term containment in a disposal facility). A thorough understanding of safety requirements facilitates the derivation of criteria addressing waste package containment and handling as well as contributing to the facility design and safety performance. However, in designing for safety, it is recognized that being overly conservative has financial implications. A more useful approach might include an evaluation of the sensitivity of modelling results to variations in key parameter values and/or reducing uncertainty regarding such parameters.

While it is good practice to use standardized packages with limited configurations, it is also good practice to allow for the flexibility to accept non-standard packages, configurations or waste types after considering the technical and safety acceptability of such non-standard packages on a case-by-case basis.

WAC need to be reasonable, concise, measurable, verifiable and appropriate to each waste stream, while providing some flexibility. It is important to note that the WAC will evolve over time. It is advisable to maintain an iterative and structured dialogue between the various operators and regulators during the entire operational lifetime of the facilities and beyond. Issues that are likely to need discussion are integration of lessons learned from the application of WAC or from departures from the recommended treatment, re-examination of the safety demonstration, new types of waste, technology advancements and so on.

Where numerical limits are stipulated in the WAC, these values need to be technically justifiable as well as quantifiable and verifiable by measurement or calculation. There needs to be a good understanding of the various measurement and calculation approaches and techniques that are available, together with their limitations and associated uncertainties.

Both the design and safety related requirements are ultimately facility specific, since they are linked to the facility design intent (e.g. types of waste to be accepted), the site location and an associated site specific safety case. Therefore, the values and limits for each of the identified key parameters can only be determined in conjunction with the design and safety assessment for a particular facility. It is important that they be based on the specific design of the intended facility. Conditions or parameter values cannot be copied from the WAC of other existing facilities (even if they are superficially very similar) to avoid the creation of inappropriate criteria.

Since the qualification of production processes and the monitoring and control of these processes are of key importance for obtaining and maintaining compliance with acceptance requirements for waste packages, it is also important for the various operators to establish their own strategies for ensuring WAC compliance in parallel with the development of the WAC. The waste generator and processor need to have an in-depth understanding of all the WAC and their significance.

WAC for all waste management steps need to be well documented, together with supporting justification. WAC documentation will also record all departures or non-conformities that were treated during the operational phase for the whole defined life of the facility (for disposal, records will be maintained after closure).

The process of improving the application of WAC will be ongoing as a result of periodic and regular reviews to identify issues and opportunities. Member States are expected to actively promote the international exchange of information on best practice and experience for mutual benefit and advancement.

Appendix I

**FULFILLING WASTE PACKAGE ACCEPTANCE CRITERIA:
SELECTION OF WASTE PACKAGES**

In the absence of a formal waste package testing programme (e.g. destructive and/or non-destructive testing), reliance will be placed on administrative control measures for selecting quality waste packages that are adequate for storage and disposal. Figures 14–19 illustrate alternatives for fulfilling WAC for waste packages based on homogeneous or heterogeneous waste [27].

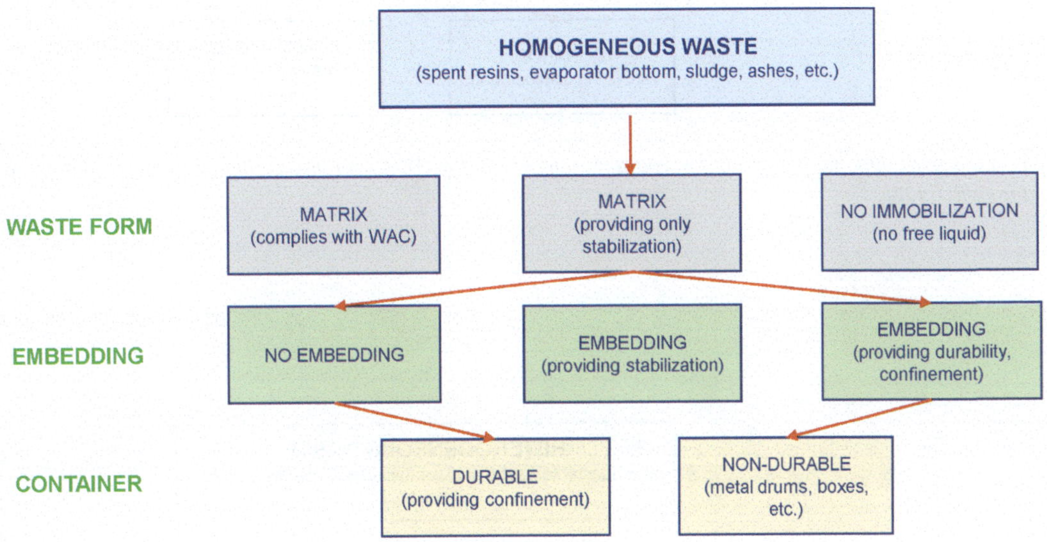

FIG. 14. Packaging alternatives for homogeneous waste non-compliant with waste acceptance criteria (WAC). Adapted from Ref. [27].

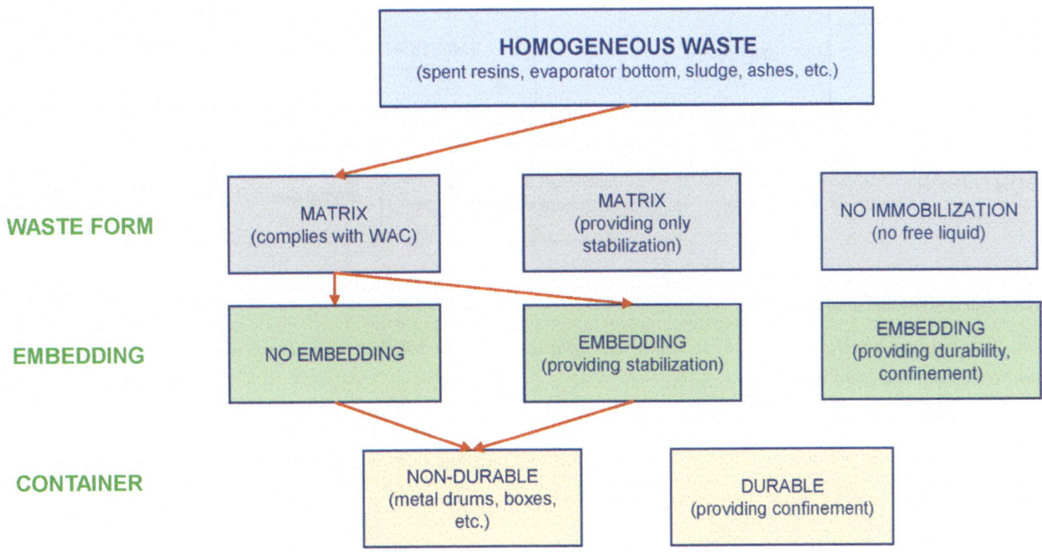

FIG. 15. Packaging alternatives for homogeneous waste complying with waste acceptance criteria (WAC). Adapted from Ref. [27].

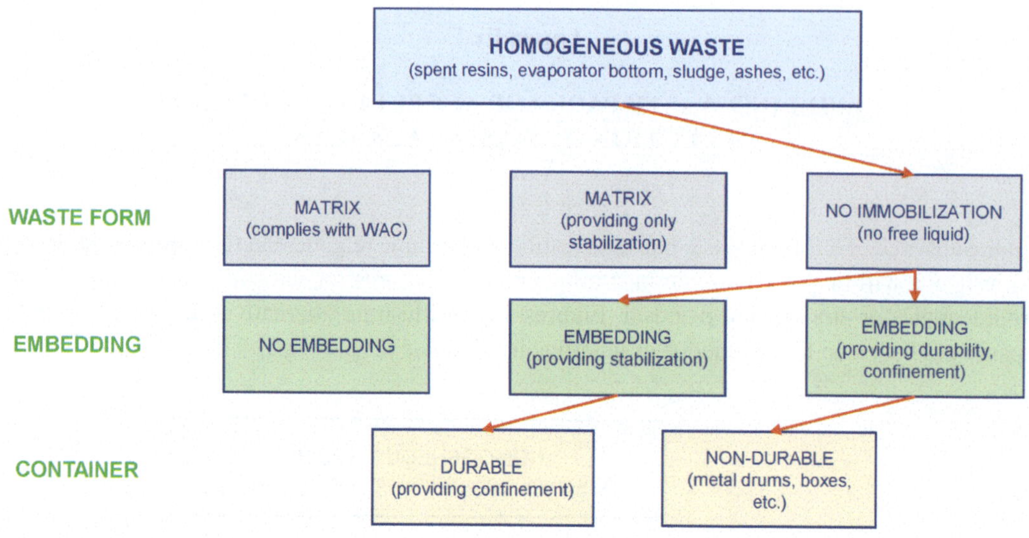

FIG. 16. Packaging alternatives to meet waste acceptance criteria (WAC) for homogeneous non-immobilized waste. Adapted from Ref. [27].

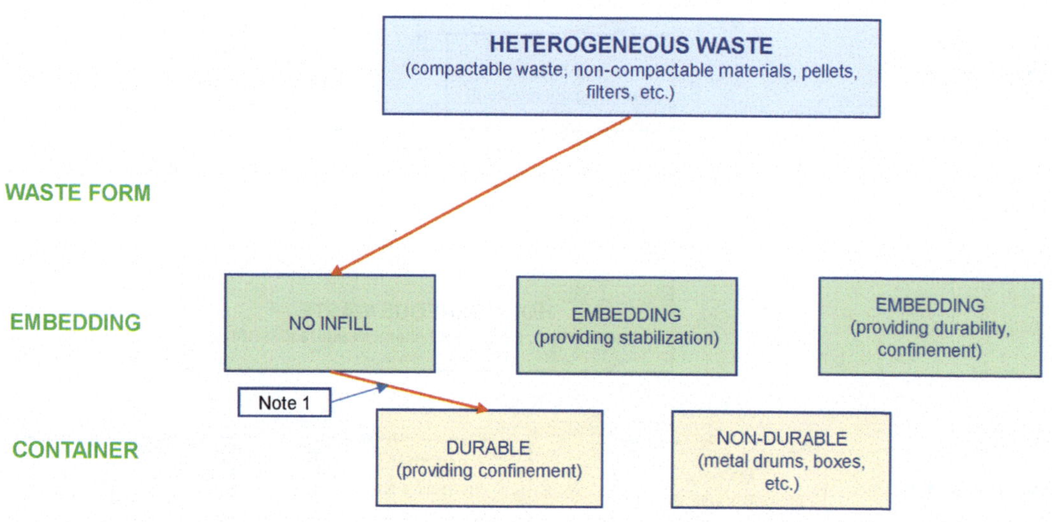

Note 1: It is advisable to fill the voids.

FIG. 17. Packaging alternatives to meet waste acceptance criteria for heterogeneous waste with no embedding or encapsulation. Adapted from Ref. [27].

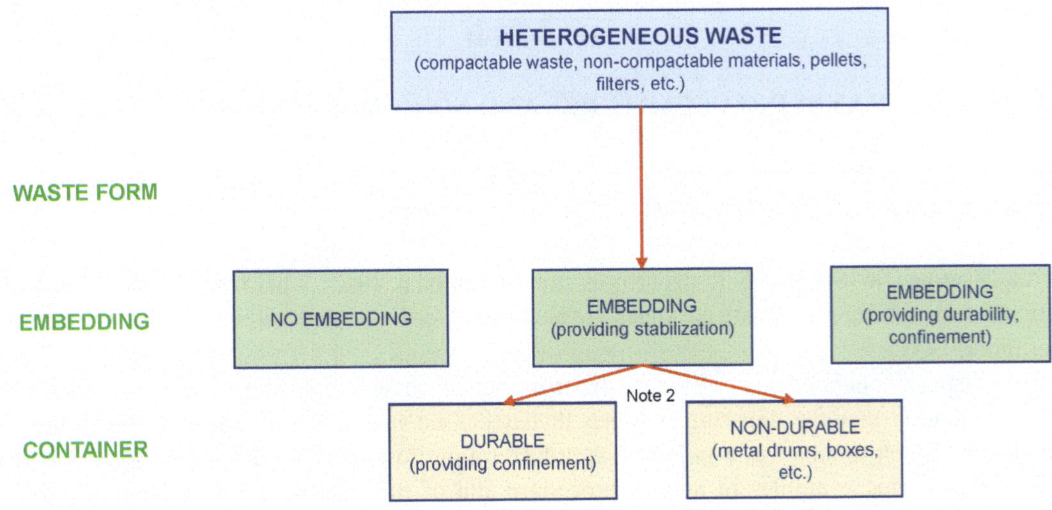

FIG. 18. *Packaging alternatives to meet waste acceptance criteria for heterogeneous waste with low quality embedding or encapsulation. Adapted from Ref. [27].*

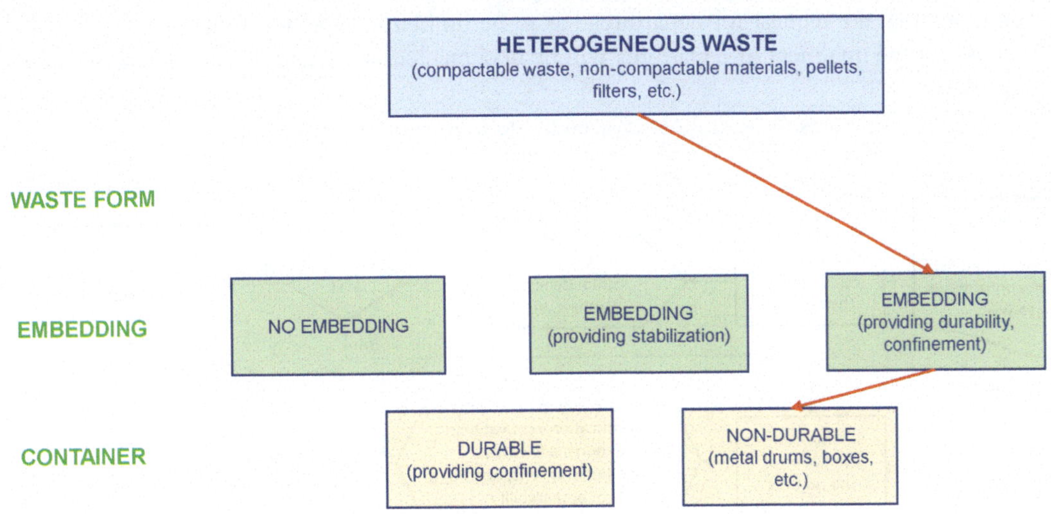

FIG. 19. *Packaging alternatives to meet waste acceptance criteria for heterogeneous waste with high quality embedding or encapsulation. Adapted from Ref. [27].*

Appendix II

HANDLING DEPARTURES AND NON-CONFORMITIES

II.1. PROCEDURE FOR THE TREATMENT OF DEPARTURES

As described in Section 5.3, departures are exceptional cases where the waste generator or conditioner realize before the waste packages are being produced that they will be unable to comply with one (or more) WAC.

The disposal operator may grant a departure under new conditions, based on justification and documentation that the departure is such that it has no negative influence on the further safe management of the primary waste package. A procedure on how to deal with these departures needs to be established and examples of such a procedure and of the minimum information required are presented in this appendix.

The content of a written document describing the methodology to follow to request a departure will include the following:

— *The purpose of the document.* Explains the objectives of the departure procedure; describes the general method to be followed if a request for a departure is introduced by a waste generator or waste processor prior to the production of the waste packages.
— *Application area.* Clearly identifies the process to be followed for application and defines the scope of the applied procedure (e.g. a request for departures can only be initiated for actual productions); highlights that the request for departures has to be limited in time and in quantity; states that the process for the treatment of departures will be documented in writing. Figure 20 shows a possible workflow.
— *Definitions.* Stipulates the terminology used in the framework of this process.

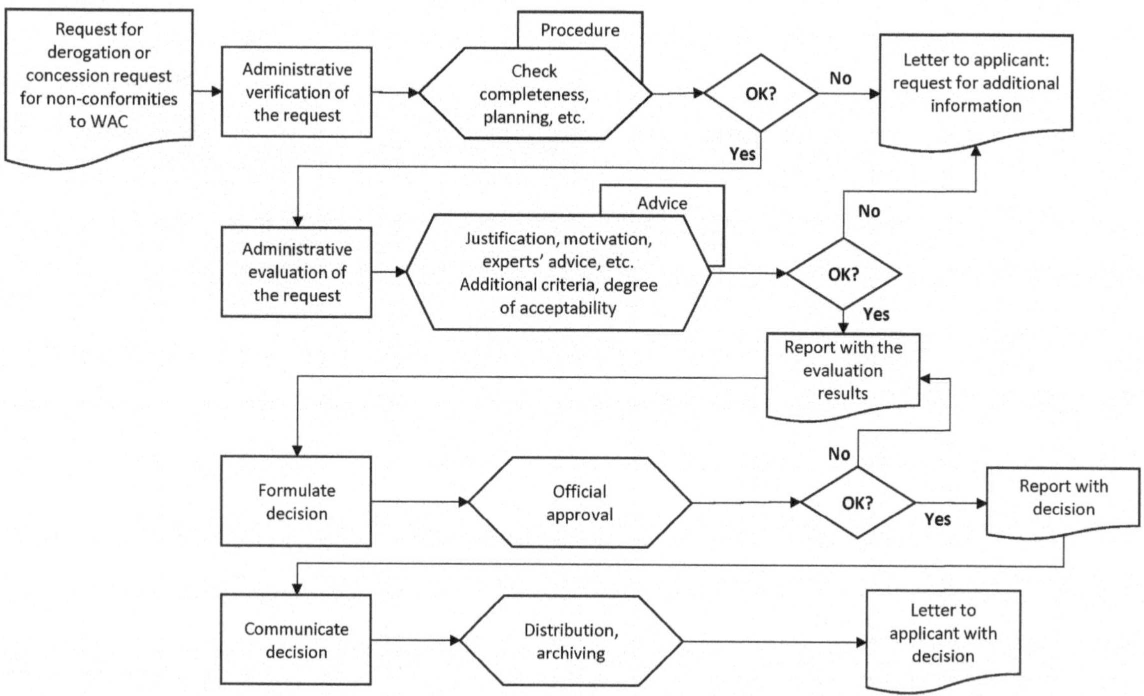

FIG. 20. Possible workflow for the treatment of a request for departures or of a non-conformity to waste acceptance criteria (WAC).

— *Responsibilities.* Defines the responsibilities of the organization or department and names the functions and responsibilities in the decision making process.

— *Principles of archiving.* Directs that the demand for departures together with the supporting documentation will be archived; establishes a procedure for archiving all associated documentation: reports from experts, reports from the decision making process, a report of the actual decision and a letter to the applicant of the request for departures.

— *References.* Provides references to directives, legislation and statutes of the organization or the department that handles the departures.

It is advisable to put into practice a process in order to provide feedback after the waste packages have been produced. The identified waste packages will be unambiguously linked with the accepted departures.

II.2. PROCEDURE FOR THE TREATMENT OF NON-CONFORMITIES

On occasions, packages may be received that, after monitoring and inspection on receipt at the waste receiver's facility, are found not to comply with the WAC. Such cases are dealt with as non-conformities. The waste receiver anticipates the possibility of non-conformities by indicating potential approaches and procedures for resolution in the receiver's WAC documentation.

A structure similar to a procedure can be applied for the treatment of non-conformities. It is advisable to first identify (e.g. by labelling, stickers, marking or tagging) all waste packages produced or to isolate these waste packages from other waste packages in order to allow unambiguous further management of non-conformities. To avoid the transfer of the waste or the paperwork prior to conclusion of treating the non-conformity, it is advisable to keep the file or documentation separate.

II.3. MINIMUM INFORMATION TO BE PROVIDED

Even though departures or non-conformities are different, similar information will be provided in order to evaluate the consequences.

(a) Identification of the waste: The concerned production (number of waste packages, volume, campaign, etc.) has to be identified. The following need to be taken into account:
  (i) In the case of a non-conformity, the waste packages will be clearly identified.
  (ii) In the case of a departure, if the exact amount of waste is not known, an estimation will be given as well as the period for which the departure will last.
  (iii) The origin of the waste has to be provided, including details on treatment and conditioning, type of (raw) waste, matrix, etc.
  (iv) The reference or link to the appropriate qualification will be given.
  (v) Quality assurance (traceability, documentation, etc.) needs to be guaranteed.
(b) Identification of WAC:
  (i) The reference to the concerned WAC needs to be given.
  (ii) All necessary details (articles, editions, versions, dates, etc.) with regard to any applicable requirement have to be listed.
  (iii) If possible, the alternative or what can be guaranteed would be addressed.
(c) Reason of non-conformity or departure:
  (i) A technical explanation would be provided.
  (ii) Parameters pertaining to the installation:
    — Radiological protection;
    — Classic security;
    — Composition and characteristics of the different constituents of the waste package.
  (iii) Financial aspects can be provided.

(d)   Scientific and technical justifications for further safe management:
  (i)   Founded justification needs to be provided.
  (ii)   All the considered alternatives, taking into account the technical and financial feasibility, will be listed.
  (iii) Scientific, technical justifications for further safe management.
  (iv)   Technical and safety aspects are necessary to justify the non-conformity.
  (v)   Results from tests, calculations and assumptions will be addressed.
  (vi)   Legislation, authorizations, directives, etc., have to be respected.
  (vii) Advice coming from security experts can be supportive and/or necessary.
  (viii) A conclusion or justification with respect to the future safe management of the waste would be provided.

Appendix III

EXAMPLES OF WASTE ACCEPTANCE CRITERIA FOR PREDISPOSAL

Processing refers to any operation that changes the characteristics of waste and could include pretreatment, treatment and conditioning [15]. The waste processor will implement an adequate QMS in a similar manner to the waste generator in order to demonstrate the fulfilment of relevant WAC to the next responsible institution in the waste life cycle.

WAC for pretreatment (any of the operations prior to waste treatment, such as collection, segregation, chemical adjustment or decontamination) will emphasize requirements linked with operations such as:

— Limits for segregation;
— Criteria for sampling;
— Restricted chemicals;
— Restricted decontamination techniques;
— Radiological characterization for difficult-to-measure nuclides;
— Conditions for chemical, physical and biological characterization;
— Selection of primary containers (drums, bags, boxes, tanks);
— Selection of materials;
— How to handle hazardous constituents;
— Considerations for further treatment or conditioning.

The following sections provide some examples of WAC for various predisposal activities in different facilities.

III.1. AQUEOUS RADIOACTIVE WASTE TO BE TREATED BY AQUA-EXPRESS-TYPE FACILITIES

The mobile modular installation Aqua-Express (Fig. 21) is a Russian Federation liquid low and intermediate level radioactive waste treatment facility designed to treat relatively small volumes of waste typically generated by research centres and other organizations. With a capacity limited to 300–500 L/h, the facility accepts aqueous waste for processing in accordance with the following WAC [28]:

— Specific activity of gamma emitters ($\Sigma\gamma$, mostly ^{134}Cs, ^{137}Cs, ^{110m}Ag, ^{140}Ba and corrosion/activation products) $\leq 1 \times 10^5$ Bq/L;
— Specific activity of beta emitters ($\Sigma\beta$, mostly ^{90}Sr) $\leq 1 \times 10^3$ Bq/L;
— Specific activity of alpha emitters ($\Sigma\alpha$) $\leq 3 \times 10^2$ Bq/L;
— pH of waste within 1–10;
— Salinity (salt content) up to 10 g/L.

III.2. SOLID RADIOACTIVE WASTE TO BE TREATED BY THE CILVA INCINERATION FACILITY IN BELGIUM

CILVA (Belgoprocess, Belgium) is a multistage incinerator that has been in operation since 1995, with a treatment capacity of 8 t/week for solid waste and up to 5 t for liquid waste [29].

FIG. 21. The Aqua-Express system. Courtesy of the Horia Hulubei National Institute for Research and Development in Physics and Nuclear Engineering, Romania.

The following radioactive waste types are accepted by CILVA for treatment:

— Raw and compacted solid waste such as personal protective clothes, gloves, rags, cotton, rubber and plastics (PVC quantity 3% average, caloric value of about 25 MJ/kg);
— Frozen animal carcasses;
— Wet or dry spent ion exchange resins;
— Organic liquid waste — scintillation liquids and organic solvents (caloric value of about 35 MJ/kg);
— Aqueous liquid waste whether or not it contains organic components and solid particles;
— Spent oil (caloric value of about 42 MJ/kg).

The maximum dose rate at the surface of each package is limited to 2 mSv/h.
The specific radioactivity of the waste is limited by:

— 40 GBq/m³ for beta-gamma emitters;
— 40 MBq/m³ for alpha emitters.

The collected ashes and fly ashes are then super-compacted and immobilized with grout.

III.3. WASTE ACCEPTANCE CRITERIA FOR STORAGE

Table 10 presents a summary of typical waste acceptance requirements for storage and related example criteria.

TABLE 10. TYPICAL WASTE ACCEPTANCE REQUIREMENTS FOR STORAGE

Type	Category	General requirement	Example criteria
Administrative	Identification	Waste/container identification and tracking	Specifications for unique labelling or package identification.
	Operational	Use of standardized containers	Specifications for use of certain package types only.
	Record keeping	Reporting requirements	Specification of format and content for documentation to accompany waste packages. Specification of duration to maintain supporting documentation.
	Quality management	Quality management requirements	Specification of a QMS or programme to be applied.
Qualification	Waste form qualification	Characterization requirements	Specification of required parameters to be characterized. Specification of acceptable methods and or tests to perform during characterization.
		Acceptable waste types/ classes/forms	Restrictions on or specification of allowable conditioning methods (e.g. cementation, polymer encapsulation). Restrictions on waste form characteristics (e.g. density, permeability, porosity, leach rate).
	Inspection, testing and monitoring	Inspection and test requirements on incoming waste packages	Specification of type and frequency of tests to be performed on wastes or waste packages.
		Periodic inspections of waste packages after disposal (until closure)	Specification of type and frequency of tests to be performed periodically after disposal of waste packages, prior to facility closure.
Technical: Design/ operational	Handling requirements	Minimum/maximum dimensions	Specification of maximum and minimum acceptable dimensions of waste packages.
		Maximum mass	Specification of maximum acceptable mass of waste packages.
		Mechanical properties of waste form/package	Specification of limits on mechanical properties such as minimum compressive strength.
		Waste package handling	Specification of standard handling equipment (e.g. fork truck) or features (e.g. lifting attachments); specification of package stacking (minimum/ maximum height, etc.); centre of gravity limits.
Technical: Safety related	Stability	Waste form chemical stability	Clear information related to compatible materials to prevent or minimize the possibility of chemical reactions between waste and container. Specific data on material properties to prevent or minimize the effects of chemical degradation over time.
		Waste form radiation stability	Specification or limit on materials to prevent or minimize possible damage to waste forms from radiolytic effects.
		Physical stability of waste form	Specification or limits on void space in waste containers.
		Durability of waste package/ container	Specification or limit on materials to ensure adequate corrosion, fire, water, mechanical impact resistance.

TABLE 10. TYPICAL WASTE ACCEPTANCE REQUIREMENTS FOR STORAGE (cont.)

Type	Category	General requirement	Example criteria
	Restrictions on content of waste packages	Restrictions on chemical or other hazardous constituents	Total ban on some materials such as toxic, explosive, pyrophoric materials. Upper limit on some materials such as corrosive, chelating materials on a per package basis.
		Restrictions on biological, pathogenic and/or infectious materials	Total ban or upper limit on a per package basis on some materials such as biological, pathogenic and/or infectious materials.
		Restrictions on free liquids	Total ban or upper limit on a per package basis of free liquids.
		Restrictions on combustible materials	Total ban or upper limit on a per package basis of combustible materials.
		Restrictions on heat generation rate	Upper limit on a per package basis of heat generation rate (e.g. decay heat of ILW).
	Radioactivity	Restrictions on radionuclides	Upper limit of activity concentration of predefined radionuclides (e.g. maximum activity concentration per waste package, per disposal cell, etc.); any requirements on distribution of radionuclides in a waste package (homogeneity of waste form).
		Restrictions on fissile content	Upper limit of concentration or total amount of fissile radionuclides.
	Radiation protection	Gamma radiation dose rates restricted levels	Upper limit of gamma dose rate at contact with and at a specified working distanced from the waste package (e.g. 1 m).
		Acceptable levels on fixed and/or removable surface contamination on waste packages	Upper limit of contamination of families of radionuclides (e.g. Bq/cm^2 of gamma, beta and/or alpha radionuclides).

WASTE CONTAINERS THAT MEET ACCEPTANCE CRITERIA

The waste package generally provides the first engineered barrier to contain radionuclides and/or hazardous chemicals during handling, storage, transportation and disposal. In achieving the required performance of the waste package, the main contributors are the waste container and the waste form. Each depends on the robustness of the other, and arguments to support this statement are as follows:

— The waste container acts as a physical barrier but also enables the waste to be handled safely during the waste package manufacturing process. A wide range of materials can be used for container manufacture, and designs are selected to comply with requirements for the packaging, transport and disposal of the wastes they contain.
— The waste form is designed to provide a significant degree of physical and/or chemical containment of the radionuclides and other hazardous materials associated with the waste. The waste form may comprise waste which has been immobilized in a medium such as cement or bitumen or which may have received simpler treatment prior to packaging (e.g. size reduction).
— The waste container can also provide additional functions such as radiation shielding and a standardized interface for handling and stacking.

Overpacks can also be used to provide additional containerization, such as when the primary container has failed, or extra shielding is required. Many containers can be used as either a primary container or as an overpack of a smaller container. Some typical waste containers and overpacks are described in Table 11.

The overall waste package can be designed to provide a number of functions during the handling, transportation, storage and/or disposal (including both the pre-closure and post-closure periods) of the waste, such as one or more of the following, depending on its role and purpose:

— Physical and/or chemical containment or control or delay of migration of contaminants;
— Shielding;
— A means for safe handling and stacking;
— Control of criticality;
— Supporting applied loads.

An important consideration in the design of a container is the length of time it is expected to perform its defined function as an engineered barrier. This is especially true for long term storage and/or disposal related functions, as illustrated in Fig. 22. A container that needs to perform for extended periods in the post-closure period will require more stringent design criteria than one that need only perform a function until the disposal facility is closed. Generally, simple containers may have expected lifetimes of several tens of years, while more highly engineered ones constructed from advanced materials may have expected design lives of several hundred years or more.

Key factors that could affect the capability of a waste package to preserve its function over the required timescale:

— The waste container design;
— The waste container materials and manufacturing processes;
— The design and technical specifications of the conditioning process;
— The rate and nature of waste form material and container degradation mechanisms;
— The interactions (e.g. their nature and rate) between the waste container and the waste form;
— The ambient environment parameters of storage and disposal facilities, such as temperature, humidity and groundwater chemistry).

TABLE 11. CHARACTERISTICS OF SOME TYPICAL CONTAINERS AND OVERPACKS

Characteristics	Type of container/overpack				
	200 L drum	200 L drum inside 400 L drum with concrete between	Concrete cont. with 200 L drum inside	Cubical concrete container	B25 box (USA)
Inner volume	200 L	200 L	200 L	1000 L	2500 L
Outer volume	200 L	400 L	1000 L	1740 L	3000 L
Dimensions (outside)	Dia. 57 cm height 88 cm	Dia. 77 cm height 110 cm	Dia. 100 cm, height 125 cm	Side length 120 cm	185 cm L × 120 cm W × 130 cm H
Loaded weight	200–500 kg		Up to 2.5 t	Up to 5 t	Up to 5 t
Wall thickness	1 mm	10 cm	20 cm	10.25 cm	2 mm
Material	Mild steel or stainless steel	Mild steel concrete	Normal concrete and mild steel	Reinforced normal concrete	Mild steel
Coating	Paint (for mild steel)	Paint	No	No	Paint
Closure	Lid with ring	Lid with ring	Sealed with concrete cap	Sealed with concrete cap	Gasketed lid
Biological shielding	None	Concrete	Concrete	Concrete	None
Strengthening	Iron bars	Iron hoops	Iron bars	Iron bars or metal fibres	Steel frame
Mechanical property	Good	Very good	Very good	Very good	Good
Easiness of handling	Good	Good	Good	Good	Good
Easiness of decontamination	Good	Good	Fair to good	Fair to good	Good
Corrosion resistant:					
In air	> 10 y	> 100 y	> 100 y	> 100 y	> 10 y
In fresh water	> 10 y	> 100 y	> 100 y	> 100 y	> 10 y
In saline water	> 10 y	> 100 y	> 100 y	> 100 y	> 10 y

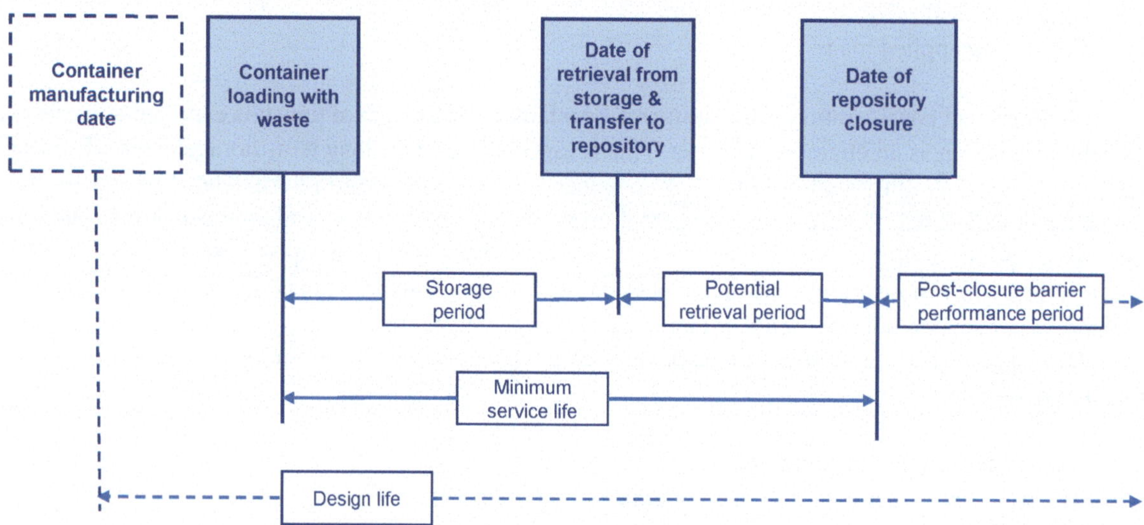

FIG. 22. Typical container life cycle.

One of the major potential threats against the ability of a waste container to maintain an adequate level of integrity for the required timescale is represented by corrosion. Other mechanisms of degradation can include the effects of heat, biodegradation, abrasion, radiolysis and chemical reactions between waste container components.

The wide range of LLW and ILW types, storage conditions and disposal environments results in a wide range of waste container materials and types. Some Member States store LLW and ILW for long periods prior to disposal in surface facilities. In other cases, generation of waste packages is closely coupled with final disposal. In addition, LLW and ILW disposal sites range from near surface facilities with and without engineered barriers in arid or humid climates, to repositories in rock formations tens to hundreds of metres underground to deep geological repositories in salt formations, to boreholes.

Examples of materials of construction for LLW and ILW disposal containers are:

— Carbon steel;
— Stainless steels, corrosion resistant alloys, duplex alloy steel;
— Concrete, reinforced concrete, and polymer impregnated concrete;
— Cast steel;
— Modular cast iron/spheroidal graphite cast iron;
— Polymers including high density polyethylene (HDPE) and fibre-reinforced composites;
— Fibre and fibre composites.

Examples and descriptions of waste containers and waste packages produced in several Member States are provided in Refs [30–33].

In selecting the proper material for the fabrication of the waste containers there is a need for the designers to:

— Understand the internal and external environmental conditions that may affect the container.
— Determine the potential degradation mechanisms (subject to the environment characteristics) and the timescales over which they could take place (to assess their impact on the storage or disposal processes).

Note that the longer the timescale (e.g. in the post-closure phase of a disposal facility), the more important slow mechanisms become. It has to be taken into consideration that all containers and waste forms will eventually degrade. They will not remain intact forever. The goal, therefore, is to design a package that will adequately retain its required properties for the required duration. Often, a conservative approach is taken in the safety assessment by not taking any credit for the waste package performance or limiting its credited duration. Additional information on containers for radioactive wastes can be found in Refs [32–34].

EXAMPLE OF A WASTE ACCEPTANCE CRITERIA DOCUMENT

Specific WAC addressing the requirements for each situation need to be developed. Several factors need to be taken into consideration by the operators or regulators when developing the WAC, such as:

— Waste types and potential waste forms;
— National standards and regulations;
— Environmental conditions;
— Specific storage/disposal parameters and characteristics.

Often, underlying specifications for the waste form and process used to produce the waste form are developed to ensure the waste form produced meets the WAC. These are supported by quality systems that ensure the waste meets these specifications. In order to generate a general WAC document that could be easily studied, the following layout is common to most WAC documents.

V.1. CHAPTER 1: GENERAL REQUIREMENTS

The section on general requirements could contain the national legislation, information on the facility design, performance assessment conclusions, site characterization, facility operational constraints, type of packages and formulating functions and responsibilities (e.g. regulator, waste operator, ownership of waste, act). For additional information, consult Section 3.

V.2. CHAPTER 2: GENERAL BASIC REQUIREMENTS ON RADIOACTIVE WASTE

Radioactive waste is generated by different kind of facilities, and all legacy, current and future radioactive waste streams, with their different chemical compositions, physical forms and radioactivity ranges, need to be classified. Coupled with this, some wastes can contain fissile materials that add criticality, safeguards and security concerns. The section on general basic requirements on radioactive waste could contain the IAEA classification (Fig. 23), or its own classification based on a safety analysis. Guidance from similar national facilities can be consulted but site specific criteria need to be approved or endorsed by the regulator. This section could also include different waste types earmarked for a disposal facility or storage facility (Fig. 24). For additional information consult Section 4 and Appendix I of this publication.

V.3. CHAPTER 3: WASTE FORM REQUIREMENTS

The section on waste form requirements could contain the specific technical requirements for storage and disposal. Typically, WAC include waste form properties, strength, leachability, container integrity, free liquids, environmental conditions (temperature, humidity, exposure to UV light, etc.), dose rates, gas generation, labelling, package tracking, concentrations, homogeneity, exclusions, package handling and stacking, as summarized in Appendix IV. The numerical limits for each requirement will be facility specific, depending on the function and importance of the waste form in the overall safety case of the facility. It is relevant to mention that some WAC limits may be derived from other considerations, such as weight, size and dose rate limits for transportation to the disposal facility, rather than from the limitations of the disposal facility itself.

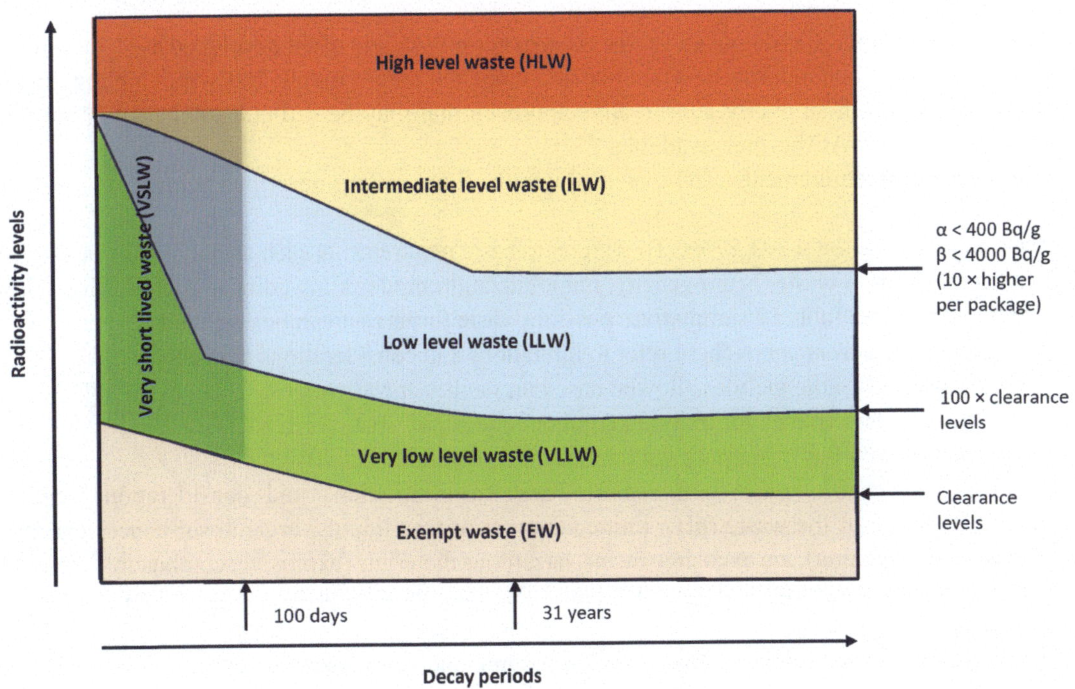

FIG. 23. *IAEA waste classification.*

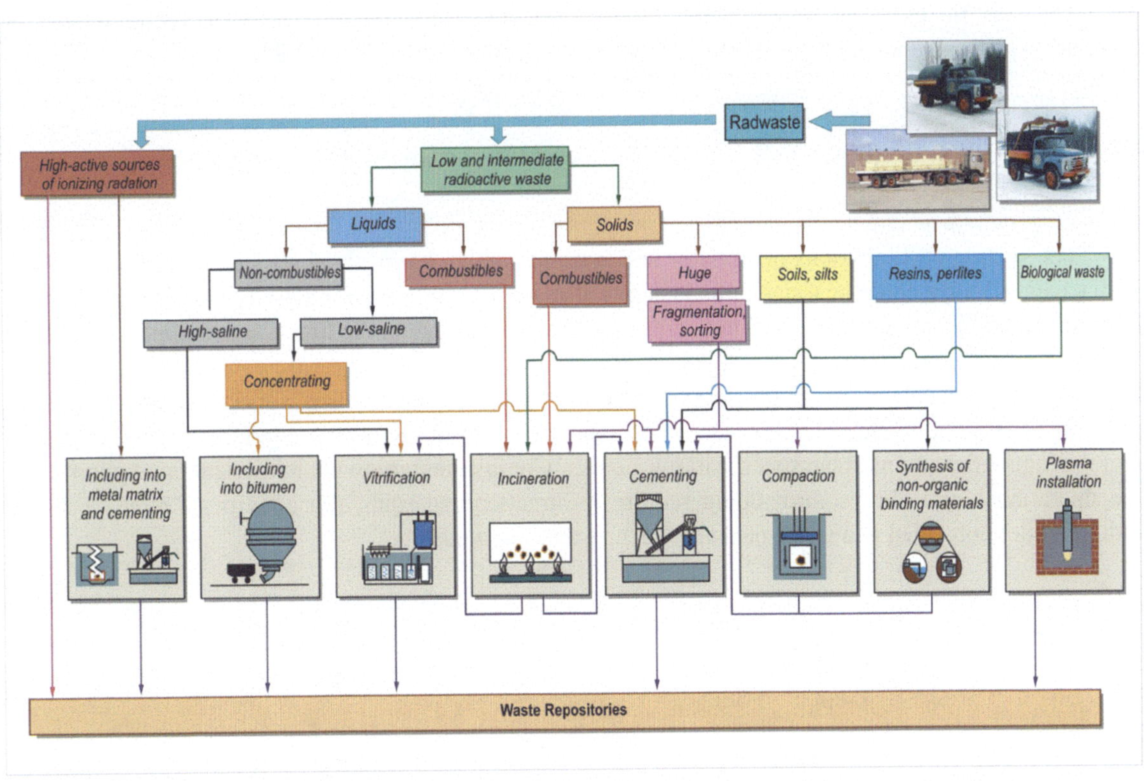

FIG. 24. *Different waste types earmarked for a disposal facility or storage facility. Courtesy of O. Karlina, Scientific and Industrial Association ('Radon'), Russian Federation.*

Where disposal facilities do not currently exist, generic WAC are often developed as an interim step. These are usually based on industry best practice or anticipated requirements. It is worth highlighting that such usage may result in an overly conservative approach that may be difficult to change in the future when facility-specific WAC become available.

The waste form requirements can be divided into the four sections described below:

(a) *Basic requirements on waste forms:* These refer to basic requirements such as only solids or solidified waste, stabilization waste, heterogeneity, immobilization binder allowed or immobilization without additional binders. Table 12 summarizes possible waste forms that can be considered.

(b) *Specific activity parameters:* These refer to limitations and considerations regarding:
   (i)   Activity and radionuclides allowed based on facility specifications;
   (ii) Permissible activities for individual radionuclides per waste form;
   (iii) Total activity per waste package (alpha and beta/gamma activity).

(c) *Specific chemical/biological parameters:* These are limitations and considerations regarding acidity/alkalinity of the waste (pH), limited amount of free liquids, organic substances (microbial degradation, chelating), ion exchange resins, hazardous materials (explosive, combustible materials), toxic chemicals (asbestos, heavy metals, benzene), material with low flash point and gas-releasing materials.

(d) *Specific physical parameters:* These are limitations and considerations regarding mass, density, heat output, compressive strength, microscopic porosity, thermal conductivity, heat capacity, radiochemical and chemical homogeneity, leach rate, swelling, etc.

The specific technical requirements summarized above contain only selected requirements for illustration purposes, and additional information can be seen in Table 13 and Table 14.

V.4. CHAPTER 4: LIMITATIONS: ACTIVITY AND NON-RADIOACTIVE HARMFUL SUBSTANCES

The section on limitations on activity and non-radioactive harmful substances could be combined with Chapter 3, or it could be a standalone chapter. The numerical limits for activity (fissile material) and non-radioactive toxic substances will be facility specific, depending on the function and importance of the waste form in the overall safety case of the facility.

The fissile content of wastes needs to be controlled to ensure that subcritical conditions are maintained under all conditions likely to be encountered at any time during conditioning and to ensure that the required waste package specifications are met. Complementary to the authorization requirements, the operational criticality control could require information regarding the activity of relevant fissile radionuclides contained in a waste package, such as the following:

^{228}Th	^{235}U	^{239}Pu	^{231}Pa	^{243}Am	^{246}Cm	^{251}Cf
^{232}U	^{236}U	^{240}Pu	^{232}Pa	^{243}Cm	^{247}Cm	^{252}Cf
^{233}U	^{237}Np	^{241}Pu	^{241}Am	^{244}Cm	^{249}Cf	^{254}Es
^{234}U	^{238}Pu	^{242}Pu	^{242m}Am	^{245}Cm	^{250}Cf	

Nuclear power plant maintenance and repair operations produce discarded equipment, organic solvents (used for degreasing and cleaning) as well as organic complexing agents from decontamination activities. These materials may contain metals such as lead, mercury, barium, chromium and nickel contaminated activated metals. The most common toxic material in evaporator concentrates is boric acid. Table 15 presents examples of the maximum concentration of contaminants classified as having toxicity characteristics that are allowed in solid waste [35].

TABLE 12. EXAMPLES OF WASTE FORMS FOR ENCAPSULATION

Waste form	Features	Limitations	Secondary waste
Glasses (vitrification)	Validated method for conditioning HLW, ILW and LLW liquids Flexibility in the range of glass formulation High reliability of the immobilization process High glass throughput High durability of the final waste form Small volume of the resulting waste form	High initial investment and operational costs Complex technology requiring high qualified personnel Generally, not economical for LLW and ILW Need to control off-gases Need to control variations in waste feed High specific energy consumption	Off-gases Filters Scrub solutions Used melters
Ceramics	Possible to incorporate higher levels of actinides than borosilicate glass Waste form is more stable and hence is more durable than glass Expected to be suitable for long term isolation since it simulates natural rocks	Limited experience; most efforts have been research based There are no known commercial installations in operation at present Generally, not considered economical for LLW and ILW The ceramic (e.g. Synroc) waste form is tailored to suit the particular characteristics of the nuclear waste to be immobilized	Filters Off-gases Scrub solutions
Glass-composite materials	Combine features of both crystalline and glassy materials Higher waste loading Higher compatibility Higher stability compared glasses	Limited experience	Off-gases Filters Scrub solutions Used melters
Cements	Widely used method for variety of LLW and ILW High flexibility Low cost Simplicity of the process Low temperature precludes volatile emissions High radiation stability, impact and fire resistance of waste forms	Increase of volume (low waste loading) Low retention of some fission and activation products Poor compatibility with organic materials and high-salt content	None
Bitumen	Mostly used for low and intermediate level waste (LILW), Chemical precipitates, low heat and low alpha wastes High flexibility High compatibility with organic materials High waste loading Low leach rate of waste forms compared with cementation	High temperature process Sensitivity to some components Low fire resistance	Filters
Metals	Extensively proven technology for conditioning of metallic waste The end product can be well categorized The end product is typically homogeneous and stable	Presorting is usually required due to dedicated melt furnaces and differences in melt temperatures of the different metals	Off-gases Slag

TABLE 13. EXAMPLES OF TECHNICAL REQUIREMENTS IN WASTE ACCEPTANCE CRITERIA DOCUMENTS FROM FACILITIES IN THE UNITED KINGDOM, SOUTH AFRICA AND SLOVAKIA

Criteria	Facility and type			
	Drigg, United Kingdom: Near surface, LLW	Vaalputs, South Africa:		Mochovce, Slovakia: Near surface, Short lived LILW
		Near surface, VLLW and LLW	Near surface, ILW	
Status	Operational	Operational	Operational	Operational
Acceptable waste forms:				
Solids	Yes	Yes	Yes	Yes
Voidage	20% allowed	Not allowed	Not allowed	5% allowed
Free standing liquid	NM	Not allowed	Not allowed	Not allowed
Sources	Allowed	Limited activity allowed	Not allowed	NM
Limits on radionuclides	Yes	Yes	Yes	Yes
Radiation external surface	2 mSv/h and 100 μSv/h	≤ 2 mSv/h	≤ 2 mSv/h	NM
Non-fixed contamination	0.4 Bq/cm^2 for all alpha 4 Bq/cm^2 for all other	0.4 Bq/cm^2 for all alpha 4 Bq/cm^2 for all other	0.4 Bq/cm^2 for all alpha 4 Bq/cm^2 for all other	NM
Other	n.a.	pH values need to be specified	pH values need to be specified	n.a.
Standard container types:				
Metal drums	Metal container with grout, of nominal density 1800 kg/m^3	Approval for use of 410 L metal drums to be obtained from regulator	Not allowed	200 L drums with concrete filling
Concrete drums	n.a.	Allowed	Approval for use of cement drums to be obtained from regulator	Standardized fibre concrete (FRC) containers
Cracks	n.a.	NM	Any cracks unacceptable	Cracks less than 50 mm long and 0.3 mm wide acceptable
Dimensional limits	NM	410 L circular metal drums	1.5 m × 1 m (diameter) circular cement drums	NM
Mass limits	NM	Up to 500 kg	Mass limit of 6 t	NM
Excluded waste types/forms:				
Aluminium	NM	Limit	Not allowed	NM
Organics	NM	Limit	Not allowed	Not allowed
Spent resin	NM	Not allowed	Allowed	NM
Inducing corrosion	NM	Limit	Not allowed	NM
Heat generating radionuclides	NM	Not allowed	Not allowed	Not allowed
Soluble solids	Allowed	Not allowed	Not allowed	NM
Biology	Not allowed	Not allowed	Not allowed	Not allowed
Gas generation	Not allowed	Not allowed	Not allowed	Not allowed
Explosive material	Not allowed	Not allowed	Not allowed	Not allowed

TABLE 13. EXAMPLES OF TECHNICAL REQUIREMENTS IN WASTE ACCEPTANCE
CRITERIA DOCUMENTS FROM FACILITIES IN THE UNITED KINGDOM, SOUTH AFRICA
AND SLOVAKIA (cont.)

Criteria	Facility and type			
	Drigg, United Kingdom: Near surface, LLW	Vaalputs, South Africa:		Mochovce, Slovakia: Near surface, Short lived LILW
		Near surface, VLLW and LLW	Near surface, ILW	
Pressured canisters	Not allowed	Not allowed	Not allowed	NM
Complexing agents	Not allowed	Not allowed	Not allowed	NM
Unconditioned ion exchange toxic material	Not allowed	Not allowed	Not allowed	NM
Hydrogen and explosive gases	NM	Not allowed	Not allowed	NM
Free standing water	Free liquid or liquids with flashpoint less than 21°C not allowed	Not allowed	Not allowed	Not allowed
Labelling	Yes, with photographic records	Yes, with photographic records	Yes	NM
References	Waste Acceptance Criteria – LLW Disposal Waste Services, WSC-WAC-LOW – Version 4.0 – March 2014	Vaalputs Waste Acceptance Criteria, JF Beyleveldt, October 2014, NECSA report no. VLP-WAC-001 rev. 07		Radioactive waste management in the Slovak Republic, Emil Bédi, June 28th, 2012

Note: n.a. — not applicable; NM — not mentioned.

V.5. CHAPTER 5: WASTE CONTAINER REQUIREMENTS

The section on waste container requirements could contain requirements regarding the physical and mechanical properties of waste containers and is based on operational conditions. Important requirements could be mass limitations, technical criteria (compressive strength, unified packages, dose rate), levels of surface contamination and surface dose rates to meet operational requirements, geometric shape and dimensions, design of internal features, lifting arrangements, materials (metal, fibre or concrete) and shape and dimensions. For additional information, consult Appendix IV.

V.6. CHAPTER 6: WASTE PACKAGE REQUIREMENTS

The section on waste package requirements could contain the general requirements for waste packages during storage before final disposal. Storage of waste packages is a very specific stage of waste management and, due to a possible long duration, it could challenge management systems.

Therefore the management systems and/or inspection procedures can be described in this section, for instance: monitoring the integrity of waste packages, controlling cooling and heating of the storage facility, maintaining operational status of equipment used in the facility, identification of waste packages, the intended movements of waste packages within the storage facility (can be performed safely, for instance, by truck), methodology to evaluate that levels of surface contamination and surface dose rates meet requirements. For additional information consult Section 4.

TABLE 14. EXAMPLES OF TECHNICAL REQUIREMENTS IN WASTE ACCEPTANCE CRITERIA DOCUMENTS FROM FACILITIES IN THE UNITED STATES OF AMERICA

	Facility and type				
Criteria	Nevada National Security Site: Near surface, LLW, ILW, mixed wastes	WIPP: Deep geological, TRU	Clive, UT: Near surface, Class A LLW	US Ecology, Richland, WA: Near surface, Class A, B, C LLW	Nevada National Security Site: Near surface, Class A, B, C LLW
Status	Operational	Operational	Operational	Operational	Operational
Acceptable waste forms:					
Solids	Yes	NM	Yes	Yes	Yes
Voidage	NM	NM	NM	NM	15% allowed
Free standing liquid	0.5% allowed	< 1 vol.% of container	1% allowed	NM	Not allowed
Sealed sources	Allowed	NM	Not allowed	NM	Allowed
Limits on radionuclides	Yes	NM	Yes	NM	50 000 curies
Radiation external surface	Packages exceeding 1 mSv/h dose rate at 30 cm to be considered for remote handling (compliance with the requirements of 10 CFR Part 835)	< 2 mSv/h at surface	< 200 mR/h on manifested container; < 500 mR/h on external, accessible surfaces of waste in container; < 80 mR/h on contact of unshielded bulk containers with resin	NM	NM
Non-fixed contamination	NM	NM	< 500 dpm α/100 cm^2; < 50 000 dpm β/100 cm^2	NM	< 2200 dpm α/100 cm^2; < 50 000 dpm β/100 cm^2
Physical properties of waste packages	Boxes measuring 4 × 4 × 7 ft or 4 × 2 × 7 ft, or 55/85 gallon drums	Type A with HEPA filter vent	Disposal containers (e.g. drums, boxes, liners), bulk containers (e.g. intermodals, gondolas) and non-bulk containers (e.g. drums, boxes) are acceptable	Class B and C waste containers will have to comply with the 10 CFR 61 and WAC 246-249 stability requirements	Standard shaped packages (square, rectangular, cylindrical) and steel drums with displacement volume of 110 gal as well as standard B-25 boxes are accepted
Radiological properties					
Limits specified	Alpha-emitting transuranic nuclides with half-lives greater than 20 years would not exceed 100 n curie (Ci)/g	> 100 nCi/g TRU; < 200 plutonium-239 fissile gram equivalent per 208 L drum	Transport vehicle (truck or railcar) not to exceed 4000 pCi/g for natural uranium or for any radionuclide in the radium-226 series, 60 000 pCi/g for thorium-230, or 6000 pCi/g for any radionuclide in the thorium decay series	NM	NM

TABLE 14. EXAMPLES OF TECHNICAL REQUIREMENTS IN WASTE ACCEPTANCE CRITERIA DOCUMENTS FROM FACILITIES IN THE UNITED STATES OF AMERICA (cont.)

Criteria	Facility and type				
	Nevada National Security Site: Near surface, LLW, ILW, mixed wastes	WIPP: Deep geological, TRU	Clive, UT: Near surface, Class A LLW	US Ecology, Richland, WA: Near surface, Class A, B, C LLW	Nevada National Security Site: Near surface, Class A, B, C LLW
Standard container types:					
Metal drums	Drums, boxes, soft-sided packages, bulk or wrapped objects allowed	n.a.	Disposal containers (e.g. drums, boxes, liners), bulk containers (e.g. intermodals, gondolas) and non-bulk containers (e.g. drums, boxes) are acceptable	n.a.	Standard shaped packages (square, rectangular, cylindrical) and steel drums with displacement volume of 110 gal as well as standard B-25 boxes are accepted
Concrete drums	Type A and B	Type A with HEPA filter vent	n.a.	Class B and C waste containers mentioned	Concrete vaults allowed
Mass limits	4082 kg (9000 lb) per box and 544 kg (1200 lb) per drum	Type A	NM	NM	Maximum mass is 24 494 kg (54 000 lb)
Labelling	Bar coded	Yes	Yes	Yes	Yes
References	Nevada National Security Site Waste Acceptance Criteria, Prepared by U.S. Department of Energy National Nuclear Security Administration Nevada Field Office, Environmental Management Operations, DOE/NV--325-Rev. 10, June 2013	Transuranic Waste Acceptance Criteria for the Waste Isolation Pilot Plant, Prepared by U.S. Department of Energy, Carlsbad Field Office, Revision 7.4, DOE/WIPP-02-3122, April 2013	EnergySolutions, Clive, UT, Bulk Waste Disposal and Treatment Facilities, Waste Acceptance Criteria Revision 9, WAC BWF-WAC-Rev-93	US Ecology Washington Facility Waste Acceptance Criteria, US Ecology Hanford Waste Acceptance Criteria	Barnwell waste management facility site disposal Criteria, B. L. Carver, ENERGY SOLUTIONS, Doc. no. S20 AD 010 REV 24

Note: n.a. — not applicable; NM — not mentioned.

TABLE 15. MAXIMUM CONCENTRATION OF TOXIC CONTAMINANTS ALLOWED IN SOLID WASTE [35]

Contaminant	Limit (mg/L)	Contaminant	Limit (mg/L)
Arsenic	5	Hexachlorobenzene	0.13
Barium	100	Hexachlorobutadiene	0.5
Benzene	0.5	Hexachloroethane	3
Cadmium	1	Lead	5
Carbon tetrachloride	0.5	Lindane	0.4
Chlordane	0.03	Mercury	0.2
Chlorobenzene	100	Methoxychlor	10
Chloroform	6	Methyl ethyl ketone	200
Chromium	5	Nitrobenzene	2
o-Cresol	200	Pentachlorophenol	100
m-Cresol	200	Pyridine	5
p-Cresol	200	Selenium	1
Cresol	200	Silver	5
2,4-Dichlorobenzene	4	Tetrachloroethylene	0.7
1,4-Dichlorobenzene	7.5	Toxaphene	0.5
1,2-Dichloroethane	0.5	Trichloroethylene	0.5
1,1-Dichloroethylene	0.7	2,4,5-Trichlorophenol	400
2,4-Dinitrotoluene	0.13	2,4,6-Trichlorophenol	2
Endrin	0.02	2,4,5-TP (Silvex)	1
Heptachlor and its epoxide	0.008	Vinyl chloride	0.2

V.7. CHAPTER 7: WASTE PACKAGE DELIVERY

The section on waste package delivery could contain the requirements regarding compliance with transport regulations, permits, marking of waste packages, management of non-conformities and departures from WAC, administrative issues (passport, unique identification, waste tracking), QA (checks and controls) for compliance with the waste package specifications and WAC, dose rate monitoring to check the contamination level, gamma and neutron monitoring to check the shielding, documentation and records (waste package specifications, waste characterization data), documentation on process knowledge, waste treatment qualification records and waste container test records (non-destructive assay, container radiography, visual examination, etc.). For additional information, consult Section 5 and the annexes in this publication.

REFERENCES

[1]   EUROPEAN ATOMIC ENERGY COMMUNITY, FOOD AND AGRICULTURE ORGANIZATION OF THE UNITED NATIONS, INTERNATIONAL ATOMIC ENERGY AGENCY, INTERNATIONAL LABOUR ORGANIZATION, INTERNATIONAL MARITIME ORGANIZATION, OECD NUCLEAR ENERGY AGENCY, PAN AMERICAN HEALTH ORGANIZATION, UNITED NATIONS ENVIRONMENT PROGRAMME, WORLD HEALTH ORGANIZATION, Fundamental Safety Principles, IAEA Safety Standards Series No. SF-1, IAEA, Vienna (2006),
https://doi.org/10.61092/iaea.hmxn-vw0a

[2]   INTERNATIONAL ATOMIC ENERGY AGENCY, Classification of Radioactive Waste, IAEA Safety Standards Series No. GSG-1, IAEA, Vienna (2009).

[3]   INTERNATIONAL ATOMIC ENERGY AGENCY, Regulations for the Safe Transport of Radioactive Material, IAEA Safety Standards Series No. SSR-6 (Rev.1), IAEA, Vienna (2018),
https://doi.org/10.61092/iaea.ur52-my9o

[4]   INTERNATIONAL ATOMIC ENERGY AGENCY, Predisposal Management of Radioactive Waste from Nuclear Power Plants and Research Reactors, IAEA Safety Standards Series No. SSG-40, IAEA, Vienna (2016).

[5]   INTERNATIONAL ATOMIC ENERGY AGENCY, Application of the Management System for Facilities and Activities, IAEA Safety Standards Series No. GS-G-3.1, IAEA, Vienna (2006).

[6]   INTERNATIONAL ATOMIC ENERGY AGENCY, Governmental, Legal and Regulatory Framework for Safety, IAEA Safety Standards Series No. GSR Part 1 (Rev. 1), IAEA, Vienna (2016).

[7]   INTERNATIONAL ATOMIC ENERGY AGENCY, Safety Assessment for Facilities and Activities, IAEA Safety Standards Series No. GSR Part 4 (Rev. 1), IAEA, Vienna (2016).

[8]   INTERNATIONAL ATOMIC ENERGY AGENCY, Predisposal Management of Radioactive Waste, IAEA Safety Standards Series No. GSR Part 5, IAEA, Vienna (2009).

[9]   INTERNATIONAL ATOMIC ENERGY AGENCY, Disposal of Radioactive Waste, IAEA Safety Standards Series No. SSR-5, IAEA, Vienna (2011).

[10]   INTERNATIONAL ATOMIC ENERGY AGENCY, The Safety Case and Safety Assessment for the Predisposal Management of Radioactive Waste, IAEA Safety Standards Series No. GSG-3, IAEA, Vienna (2013).

[11]   INTERNATIONAL ATOMIC ENERGY AGENCY, Storage of Radioactive Waste, IAEA Safety Standards Series No. WS-G-6.1, IAEA, Vienna (2006).

[12]   INTERNATIONAL ATOMIC ENERGY AGENCY, The Safety Case and Safety Assessment for the Disposal of Radioactive Waste, IAEA Safety Standards Series No. SSG-23, IAEA, Vienna (2012).

[13]   INTERNATIONAL ATOMIC ENERGY AGENCY, Near Surface Disposal Facilities for Radioactive Waste, IAEA Safety Standards Series No. SSG-29, IAEA, Vienna (2014).

[14]   Joint Convention on the Safety of Spent Fuel Management and on the Safety of Radioactive Waste Management, INFCIRC/546, IAEA, Vienna (1997).

[15]   INTERNATIONAL ATOMIC ENERGY AGENCY, IAEA Nuclear Safety and Security Glossary, 2022 (Interim) Edition, IAEA, Vienna (2022),
https://doi.org/10.61092/iaea.rrxi-t56z

[16]   INTERNATIONAL ATOMIC ENERGY AGENCY, Leadership, Management and Culture for Safety in Radioactive Waste Management, IAEA Safety Standards Series No. GSG-16, IAEA, Vienna (2022).

[17]   INTERNATIONAL ATOMIC ENERGY AGENCY, Strategy and Methodology for Radioactive Waste Characterization, IAEA-TECDOC-1537, IAEA, Vienna (2007).

[18]   INTERNATIONAL ATOMIC ENERGY AGENCY, Categorizing Operational Radioactive Wastes, IAEA-TECDOC-1538, IAEA, Vienna (2007).

[19]   INTERNATIONAL ATOMIC ENERGY AGENCY, Retrieval, Restoration and Maintenance of Old Radioactive Waste Inventory Records, IAEA-TECDOC-1548, IAEA, Vienna (2007).

[20]   INTERNATIONAL ATOMIC ENERGY AGENCY, Methods for Maintaining a Record of Waste Packages During Waste Processing and Storage, Technical Reports Series No. 434, IAEA, Vienna (2005).

[21]   INTERNATIONAL ORGANIZATION FOR STANDARDIZATION, Quality Management Systems – Requirements, ISO 9001:2015, ISO, Geneva (2015).

[22]   INTERNATIONAL ATOMIC ENERGY AGENCY, Development of Specifications for Radioactive Waste Packages, IAEA-TECDOC-1515, IAEA, Vienna (2006).

[23]   INTERNATIONAL ATOMIC ENERGY AGENCY, Derivation of Activity Limits for the Disposal of Radioactive Waste in Near Surface Disposal Facilities, IAEA-TECDOC-1380, IAEA, Vienna (2003).

[24]   AUSTRALIAN GOVERNMENT, Preliminary Safety and Waste Acceptance Report of the National Radioactive Waste Management Facility (NRWMF) (2018),
https://www.industry.gov.au/sites/default/files/2019-09/nrwmf-preliminary-safety-and-waste-acceptance-report.pdf

[25]   TORRES, P., CAHEN, B., DUTZER, M., "Low and Intermediate Level Short Lived Waste Disposal in France: Improving Confidence", paper presented at Waste Management Symp. 2015, Phoenix, AZ, 2015.

[26]   INTERNATIONAL ATOMIC ENERGY AGENCY, Inspection and Verification of Waste Packages for Near Surface Disposal, IAEA-TECDOC-1129, IAEA, Vienna (1999).

[27]   OJOVAN, M.I., "The role of Waste form in Complying Storage/Disposal Waste Acceptance Criteria," paper presented at the Regional Workshop on Waste Acceptance Criteria Development and Use, Bucharest, 2016.

[28]   KARLIN, Y.V., DMITRIEV, S.A., ILJIN, V.A., OJOVAN, M.I., BURCL, R. "Elaboration of not large mobile modular installations 'Aqua-Express' (300 L/h) for LWR cleaning", Paper WM03-120, Proc. Waste Management Conf. 2013, 23-27 February 2003, Tucson, AZ (2003).

[29]   DECKERS, J., "Incineration and plasma processes and technology for treatment and conditioning of radioactive waste", Ojovan, M.I., Handbook of Advanced Radioactive Waste Conditioning Technologies, Woodhead, Oxford (2011) 43–66.

[30]   INTERNATIONAL ATOMIC ENERGY AGENCY, Containers for Packaging of Solid Low and Intermediate Level Radioactive Wastes, Technical Reports Series No. 355, IAEA, Vienna (1993).

[31]   INTERNATIONAL ATOMIC ENERGY AGENCY, Interim Storage of Radioactive Waste Packages, Technical Reports Series No. 390, IAEA, Vienna (1998).

[32]   INTERNATIONAL ATOMIC ENERGY AGENCY, Evaluation of Low and Intermediate Level Radioactive Solidified Waste Forms and Packages, IAEA-TECDOC-568, IAEA, Vienna (1990).

[33]   NUCLEAR DECOMMISSIONING AUTHORITY, Guidance on the Application of the Waste Package Specifications for Shielded waste packages, Report WPS/702/01, UK NDA, Harwell (2014).

[34]   NUCLEAR DECOMMISSIONING AUTHORITY, Guidance on the Application of the Waste Package Specifications for unshielded waste packages, Report WPS/701/01, UK NDA, Harwell (2014).

[35]   NUCLEAR REGULATORY COMMISSION, Identification and Listing of Hazardous Waste under Title 40: Protection of Environment, Part 261, of the U.S. Code of Federal Regulations, current as of April 16, 2014.

Annex I

WASTE ACCEPTANCE CRITERIA DEVELOPMENT
METHODOLOGY: THE APPROACH IN BELGIUM

I–1. PROCESS OF ACCEPTANCE

The waste acceptance system applied by the Belgian Agency for Radioactive Waste and Enriched Fissile Materials (ONDRAF/NIRAS) consists of three parts: the acceptance criteria, a qualification process and an acceptance process. Figure I–1 shows that every stage of the acceptance of radioactive waste is accompanied by the waste acceptance criteria (WAC).

I–2. WASTE ACCEPTANCE CRITERIA

The WAC are stated in formal documents (referred to as ACRIAs) and they represent the minimum requirements applicable for each waste category in order to be accepted for a given management step. The WAC are accompanied by technical notes and procedures that set the level of details of specific technical aspects and acceptance arrangements.

I–3. QUALIFICATION OR AGREEMENT

The technical qualification dossier for both conditioned or unconditioned waste is prepared and submitted by the producer. The technical qualification dossier, which is a necessary condition of waste acceptance, contains the demonstration of suitability of methods and equipment, as well as documentation of the on-site inspections and the producers' quality system by ONDRAF/NIRAS.

The objective of the qualification process is to confirm that the characterization methodologies used, the whole treatment and conditioning technical infrastructure and the primary waste package meet all the requirements that are mandatory to produce a final waste package which complies with the applicable WAC. The WAC represents the guiding principle for the qualification process that has the purpose to ensure that a method, process or facility will produce final waste forms and packages that meet the requested acceptance criteria. Only those radioactive wastes that pass the qualification process are accepted by ONDRAF/NIRAS; the qualification process contributes to guaranteeing that all safety rules and conditions are fulfilled during the waste management steps.

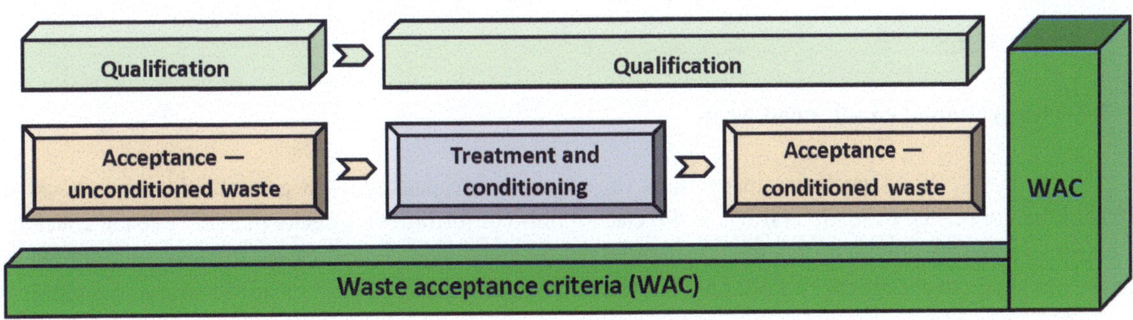

FIG. I–1. Radioactive waste management system principles.

I–3.1. The waste processing, conditioning and storage facilities

There are three stages that need to be accomplished for qualification:

(1) The technical qualification file completed by the waste generator have to be approved by ONDRAF/ NIRAS. In the technical qualification file, the waste generator describes the process and facility functioning as well as their ability to deliver end products which are compliant with the applicable acceptance criteria.
(2) The verification (technical audit), at the facility site, of the conformity between the information provided by the waste generator in the technical qualification file and its implementation in the facility. The report of the technical audit is subject to ONDRAF/NIRAS approval.
(3) The documentation prepared by the waste generator to prove the conformity of the waste produced and verified in the facilities. The compliance documentation is subject to ONDRAF/NIRAS assessment and approval.

I–3.2. Waste characterization methodology and ensuring compliance with acceptance criteria

The radiological characterization methodology qualification is also based on specific qualification files. The same approach could be applied for qualifying the measuring facility/equipment and method additionally to the methods in place to demonstrate that the waste meets the acceptance criteria.

I–3.3. Conditioned waste primary package

The qualification process for the primary package for conditioned radioactive consists of checking its compliance with the ONDRAF/NIRAS specific requirements throughout the combined inspection and documentation verification. The qualification process role is to determine a priori the conformity of the waste with applicable WAC.

I–4. TREATMENT OF DEPARTURES

If a waste generator or waste processor is unable to comply with one (or more) WAC, a departure may be granted by ONDRAF/NIRAS. To apply for a departure, the waste generator or waste conditioner need to prove that this departure has no negative influence on the further safe management of the primary package. For treating departures, ONDRAF/NIRAS applies a procedure of the Review Committee for Acceptance Criteria (RCA).

I–5. CONTROLS

1–5.1. Inspection of conditioned waste

ONDRAF/NIRAS performs inspections on certain waste packages. In general, this is a check of 10% of the concerned campaign (100% in the case of non-conformities). Checks include a visual check of the drum's integrity and identification, verification of the mass, and control of the surface contamination. These physical checks are carried out on the site of the waste generator. Results are recorded in a report.

After acceptance, the waste is transported to the centralized storage buildings at the Belgoprocess site (operated by a subsidiary of ONDRAF/NIRAS). Before they are taken into the storage building, Belgoprocess checks the identification and integrity of each drum. The results are recorded in a report. The drums are inspected by camera and pictures are taken for later comparison. For each intake a reference drum is selected, its surface contamination is checked and its mass and dose rate are measured. In some cases, this drum can be the reference drum in the monitoring programme of accepted conditioned waste.

I–5.2. Monitoring programme of accepted conditioned waste

In order to monitor characteristics of the radioactive wastes that are to be transferred to ONDRAF/NIRAS, several administrative and technical procedures were developed. The main scope of these procedures is to periodically confirm that the wastes that have been accepted are still compliant with the applicable WAC and that they are still compatible with the reference solution considered for their long term management.

Regarding the monitoring of accepted conditioned waste packages, ONDRAF/NIRAS General Rules stipulate:

(a) The administrative and technical procedures for the follow-up of the characteristics and properties of the conditioned waste packages are established by ONDRAF/NIRAS. The double goal of these procedures is to verify that the primary packages:
  (i) Are still compliant with the applicable WAC at the moment of signing the acceptance report;
  (ii) Remain compatible with their reference final destination (host rock, storage and disposal concept, etc.).
(b) The frequency of this follow-up will be as follows:
  (i) The first check will occur at three years after the acceptance of the primary package.
  (ii) The subsequent checks are planned at every ten years during the storage phase.

ONDRAF/NIRAS is documenting the results of these check in a technical report which will be presented to and signed by the waste producer/generator.

This programme is only for accepted conditioned waste packages. Reference packages (representative for one campaign of accepted waste) are chosen based on the nature of waste that was conditioned and the type of container used.

The objective of this long term monitoring is to verify if the conditioned waste packages remain in conformity with the acceptance criteria that were applicable when the protocol of acceptance was established, and to ensure they remain compatible with the requirements of their reference final destination. If the acceptance criteria are modified, for example as a consequence of an important evolution of the reference final solution, it could be necessary to monitor or measure other characteristics of the waste packages, either during the storage period, or just before disposal.

I–6. COMPLIANCE VERIFICATION

I–6.1. Purpose of acceptance of conditioned waste packages

Once the radioactive waste is treated and conditioned, the final conditioned waste package needs to be accepted for transport, storage and/or disposal. The acceptance process outcomes will guarantee that is sufficient knowledge concerning the final product and that the waste package characteristics are fully compliant with the well defined waste package specifications.

The objectives of the radioactive waste acceptance process are twofold:

(a) To ensure that ONDRAF/NIRAS is in a position to fulfil all its statutory tasks laid down in the legislation (transfer of responsibility of management);

(b) To liberate the necessary funds to cover the operating costs and the provisional costs (of ONDRAF/NIRAS and of its subcontractors) through each envisioned step in the safe management of the radioactive waste.

Which are the steps in the acceptance of a conditioned waste package?

(1) The waste generator initiates the acceptance procedure by sending to ONDRAF/NIRAS an application for acceptance of radioactive waste. The waste generator will fill a production documentation file that (in line with the applicable WAC and the approved qualification file) will contain the required information regarding the waste.
(2) The acceptance is based on the verification of the compliance of the production file with the applicable WAC for the conditioned waste package.
(3) The documentation file is controlled by ONDRAF/NIRAS, who will certify the documented quality of a conditioned waste package that was produced in a qualified conditioning facility.
(4) In addition, physical controls are carried out as part of the acceptance procedure to check the conformity of certain documented values declared by the waste generator.
(5) During the acceptance process the verification of the radiological and (for certain waste types) chemical waste characteristics are performed, to validate the individual data for each waste package produced during the conditioning campaign. This verification relies on the notes of methodology for the radiological and chemical characterization of the conditioned waste package as compared with theoretical technical references.
(6) The data related to the homogeneity of the conditioned waste package is also checked during the acceptance process, based on different official sources of data (documents, databases, application forms).

All these activities are documented and result in an official correspondence with the waste generator. A consolidated documentation file, which includes the waste generator production documentation file and the documentation generated by ONDRAF/NIRAS, is further managed for periodic surveillance during the storage of the conditioned waste package by ONDRAF/NIRAS. Each documentation file (with a unique ONDRAF/NIRAS reference) contains references to, or copies of, the applicable WAC, the approved qualification file, the methods for characterization and different types of reports related to inspections, administrative controls, acceptance, transfer, storage, monitoring of the radioactive waste, etc. The waste package information is summarized in an official ONDRAF/NIRAS document (application for acceptance) for unconditioned waste and conditioned waste. The documentation structure is summarized in Fig. I–2.

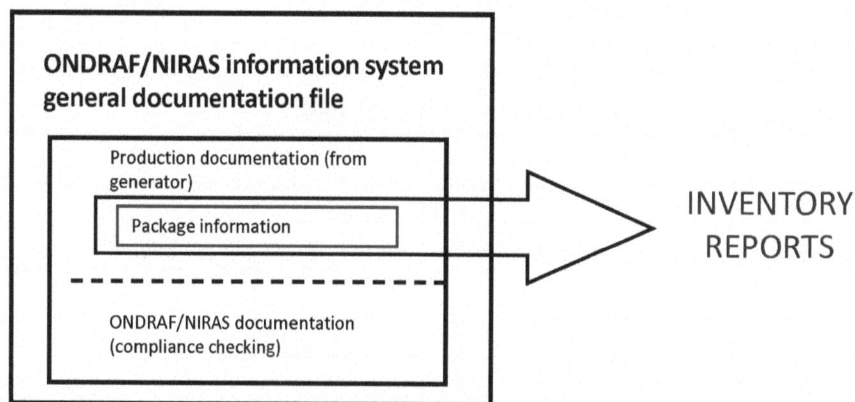

FIG. I–2. Structure of the Belgian Agency for Radioactive Waste and Enriched Fissile Materials (ONDRAF/NIRAS) documentation.

I–7. TREATMENT OF NON-CONFORMITIES

In the case of non-conformities, ONDRAF/NIRAS can either refuse the primary package and require the conditioner to take the necessary corrective measures so that the primary package complies with the WAC or accept the primary package without corrective measures, provided proof is given that the non-conformity has no negative influence on the further safe management of the primary package. In this case, it is up to the producer or conditioner to provide this demonstration. Non-conformities are treated by ONDRAF/NIRAS through the RCA.

Treatment of non-conformities is done in consultation with ONDRAF/NIRAS and the operators of the treatment, conditioning and storage facilities. Non-conformities are treated on a case-by-case basis. Four categories for the treatment of non-conformities (with regard to long term criteria) for conditioned waste packages are possible:

— *Category 1: Unacceptable non-conformities.* A conditioned waste package is not acceptable, and the waste generator has to take corrective actions so that the conditioned waste package complies with the WAC.
— *Category 2: Non-conformities that could be accepted.* It is the waste generator's responsibility to provide demonstration of the packages' acceptability for storage and eventual disposal. A request can be lodged by the waste generator, ONDRAF/NIRAS or the operator of the storage facility. The demonstration provided by the waste generator will be evaluated by the RCA. The package will then be given either a positive judgement — waste is acceptable as an exception with additional conditions (technical and/or financial) — or a negative judgement — waste is not acceptable.
— *Category 3: Non-conformities with declassification.* The conditioned waste package does not comply with the proposed WAC but, after reclassification to another waste class or category, it can comply with another WAC.
— *Category 4: Non-conformities without consequence.* A conditioned waste package can be accepted but the non-conformity will have to be documented. If this is repeated, the waste generator will be requested to undertake corrective actions.

I–8. METHODOLOGY TO TRANSFER WASTE

ONDRAF/NIRAS is a public organization which is responsible for the management of all radioactive wastes in Belgium, providing a complete set of services: collection/transportation from the generator site, treatment, conditioning, storage and future disposal [I–1].

REFERENCE TO ANNEX I

[I–1] DE BOCK, C., "Preliminary Waste Acceptance Criteria for the Surface Disposal Facility in Dessel (BE)", presented at Workshop on Waste Acceptance Criteria for Disposal of (V)LILW, Peine, Germany, 2010.

Annex II

**WASTE ACCEPTANCE CRITERIA DEVELOPMENT
METHODOLOGY: THE APPROACH IN FRANCE**

II–1. ALLOCATION OF RESPONSIBILITIES

In France, the interface between the different waste management entities — i.e. the waste generator, the National Radioactive Waste Management Agency (Andra), the waste management organization and the Nuclear Safety Authority — is of primary importance [II–1 to II–3]. The arrangement is illustrated in Fig. II–1, which shows, in particular, the place of the WAC within the process. At each stage of the process from agreement to the reception at the Centre de l'Aube disposal facility, the question to be answered is: 'Do the waste packages comply with the requirements?'

II–2. THE WASTE ACCEPTANCE CRITERIA

The objective is to ensure that the waste package complies with the WAC. This is audited by Andra through an agreement process where Andra obtains confidence in the ability of the waste generator to produce waste packages as well as conducting the surveillance and control processes in order to maintain confidence.

All the operational parameters are grouped in a list, entitled the Contractual Requirements Sheet, which contain the package description following closely Andra's requirements. Basically, it contains all the package specification details that the generator agrees to produce and that Andra agrees to accept for disposal.

The waste package criteria depend on normal and accidental scenarios. Maximum weight, maximum dose rate and external contamination requirements are driven by normal operation and radiological protection needs. Fire resistance and drop resistance are derived from accidental scenarios considered in the safety analysis at the Centre de l'Aube.

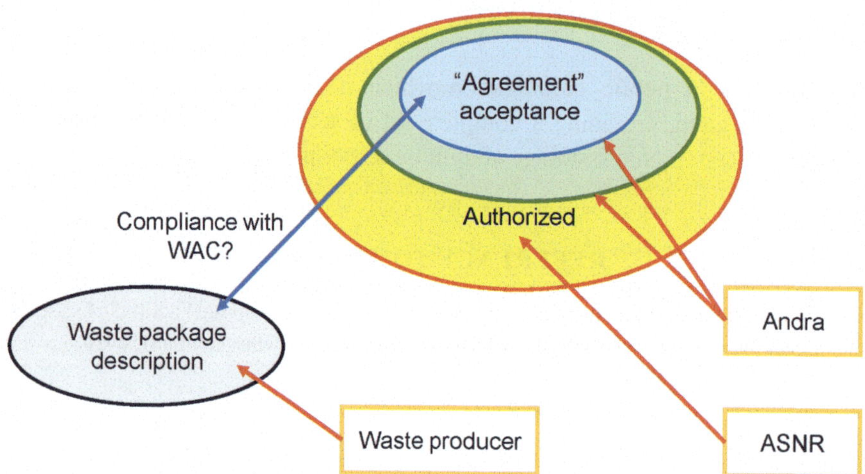

FIG. II–1. Illustration of responsibilities. Adapted from Ref. [II–1]. Andra — the French National Radioactive Waste Management Agency; ASNR — French Nuclear Safety and Radiation Protection Authority; WAC — waste acceptance criteria.

II–3. THE AGREEMENT PROCESS

According to the French basic safety rule BSR III.2.e, an agreement has to be prepared relating to a 'family' of waste packages prior to any of these waste packages being delivered. A family of packages will employ the same process with the same container type and will have the same characteristics. In that way, the agreement process leads to an operational definition of the package that complies with Andra's requirements and to relevant quality assurance procedures (see Fig. 12, Section 5.2).

The agreement states that Andra considers that packages of this family will comply with the disposal facility requirements and are approved by Andra after a satisfactory review of the documentation provided by the waste generator. The documentation will include the process description, the radionuclide characterization method, the waste package description and quality assurance measures. Acceptance of a waste package at the Andra disposal site requires that it complies with the agreement.

II–4. THE IMPORTANCE OF THE SURVEILLANCE AND CONTROL PROCESS

Some parameters specified by Andra may be monitored directly after the production of the waste package or during the process of production. For instance, the weight of a package can be readily measured and compared with the maximum specified weight. Similarly, the dose rate can be measured and compared. In other cases, it is not possible to measure certain containment parameters (leaching rate, diffusion coefficient, etc.) directly during the fabrication process of the package. For these parameters, it would be necessary to destroy the package to obtain the measurement.

Various methods are implemented for surveillance and control such as:

(a) Computer checking of the package properties declared by the waste generator.
(b) Control of packages on delivery to the Centre de l'Aube facility, e.g. measurements are made of the dose rate and surface contamination of packages.
(c) Audits performed in the facility of the waste generator and/or processor.
(d) Destructive examination (core samples or total cutting) and non-destructive tests (weight, dimensional controls, dose rate, surface contamination, gamma spectrometry, etc.) on actual packages. About two hundred non-destructive tests a year and 10 to 20 destructive tests a year are performed.

II–5. THE MANAGEMENT OF NON-CONFORMITIES

Different types of non-conformities can occur:

— During the agreement process, the waste generator may submit to Andra a request to deviate from one part of the WAC (e.g. to use a different technical analysis method from that which was recommended). In this case, the whole production of the family concerned is affected and may have consequences for the number of waste packages produced.
— If one or several packages already produced do not meet the WAC, the cause of the non-conformity will be described by the producer. For example, details of the malfunctioning process or handling problem.

When waste packages do not comply within the domains of the agreement, they need to be checked to see if they still comply with the WAC. If not, a particular safety analysis is needed to confirm that the waste package still meets the authorized domain. Depending on the safety assessment, the result may lead to three outcomes: accepted, refused or accepted under condition. The treatment of non-conformities can lead to modification of the WAC and/or the operational process.

REFERENCES TO ANNEX II

[II–1]   TISON, J.L., "Waste Management Acceptance, Criteria and Waste Categorization, related to the French Organization, DG/AI JLT 04-06 CEG, presented in Stockholm, 2006.

[II–2]   FINSTER, M., KAMBOJ, S., International Low-Level Waste Disposal Practices and Facilities, prepared for U.S. Department of Energy Used Fuel Disposition Campaign, Argonne Natl Lab, October 7, 2011, ANL-FCT-324, Argonne Natl Lab, Lemont, IL (2007),
https://doi.org/10.2172/1033484

[II–3]   VOINIS, S., MAILLARD J.L., "Waste Acceptance Criteria for LLW, Centre de l'Aube", presented at IAEA TC Regional Workshop, Vilnius, 2009.

Annex III

**WASTE ACCEPTANCE CRITERIA DEVELOPMENT
METHODOLOGY: THE APPROACH IN SLOVAKIA**

III–1. PROCESS FOR DEVELOPMENT OF WASTE ACCEPTANCE CRITERIA

According to §2 (f) of the Slovak Atomic Energy Act No. 541/2004 Coll., nuclear waste repositories are considered nuclear installations and, in order to obtain a licence for construction and operation of a repository, the applicant will have to submit the document "Limits and Conditions (L&C) of Safety Performance". This document is then approved by the national Nuclear Regulatory Authority (NRA) during the licensing process [III–1].

NRA Regulation No. 53/2006 Coll., which regulates radioactive waste management and spent fuel management, states:

— Only package forms that comply with the L&C of safe operation of facilities, approved by the NRA and based on a safety analysis, are acceptable.
— The content of the L&C document will be based on the safety analysis. The components of the L&C (acceptance criteria) for safety operations for the repository specified are:
  - Waste package type and structural stability;
  - Activity content of relevant radionuclides;
  - Leachability;
  - Thermal and radiation impacts;
  - Possibility of critical state formation or possibility of microbial degradation;
  - Gas generation;
  - Corroding, explosive and self-flammable substances;
  - Combustible substances;
  - Free liquids and complexing agents;
  - Surface contamination, dose rate;
  - Dimensions, mass and labelling.

The L&C document is divided into three sections: safety limits, limits and conditions of safe operation, and administrative management. Each section contains:

— Objectives – define the purpose of limit condition;
— One or two conditions for parameter or facility state;
— Regime definition (containers are disposed of, containers are not disposed of);
— Response of personnel in the event of L&C failure;
— Requirements on checking (periodicity, kind and range).

The approved L&C of the Mochovce disposal facility contain definitions of the acceptable forms of waste packages. Sets of disposal waste packages have been approved by the regulatory authority based on the safety analyses.

III–2. PROCESS OF ACCEPTANCE

III–2.1. Qualification or agreement

Technology licensing by the regulatory body ensures that waste packages produced comply with the repository L&C. Technology licensing includes:

— Waste container manufacturing;
— The operation of waste treatment technologies;
— The handling and transport of waste packages;
— The operation of the repository.

The waste treatment operator and the repository operator have operational instructions describing:

— Radiation inspections;
— Technological regulations;
— The internal quality assurance programme;
— Instructional operations regulation for the supervision of facilities;
— Inspection of the radioactive waste conditioning and packaging at the producer site;
— Visual inspection of waste packages;
— Transport of waste packages;
— Disposal of waste packages into the repository.

Internal and external monitoring includes:

— Regular control and analysis of the final package by the waste treatment operator;
— Periodic inspection by the repository operator on the waste treatment or producer site;
— Periodic inspection by the regulatory body;
— Occasional audits by an internal or an external auditor.

III–2.2. Treatment of departures

Some foreseeable and common departures have acceptable limits and contingencies already defined in the L&C. Treatment of some other departures, not precisely defined in the repository L&C, are dealt with on a case-by-case basis. For example, according to the L&C it is acceptable to have small defects (cracks) on the surface of concrete waste packages. The width of these cracks will be less than 0.3 mm and the length of these cracks will be less than 50 mm. In individual cases it is accepted that the length of cracks on the surface can be higher than 50 mm (but the width of cracks has to be less than 0.3 mm). These are the so-called reduced integrity fibre-concrete containers. The number of reduced integrity fibre-concrete containers that may be disposed of at the repository is strictly limited according to the safety report.

III–2.3. Controls

Verification of waste package characteristics includes:

— Chemical and radiochemical analysis of the input waste and other input components.
— Chemical and radiochemical waste treatment process controls.
— Waste package monitoring (surface dose rate and contamination measurement) and non-destructive assay of waste packages by gamma scanning and passive neutron counting.
— Waste stream monitoring performed by the waste generator.

— Declaration of the radionuclide content in the waste package before its disposal. This is based on the measurement of ^{60}Co and ^{137}Cs nuclides in the representative sample of raw waste streams and conditioned waste.
— Calculation and optimization of the type and amount of radioactive waste within the final waste package.

Waste monitoring results are compared with approved limits and the consignment declaration to confirm conformance.

III–2.4. Compliance verification

Before transport of the waste package from the waste treatment site to the repository:

— All relevant documentation needs to be prepared and all significant parameters (radiation protection, chemistry, radionuclides contents, physical parameters of FCC and so on) are to be controlled by staff responsible for radioactive waste treatment.
— The waste package documentation is controlled by the repository operator on the waste treatment site.
— The physical control (security) of waste packages will be assured.
— The preliminary approval of the waste package will be agreed between the waste producer and the repository operator.

Upon delivery to the repository and before disposal of the waste package:

— All relevant documentation and all significant parameters of the waste packages are checked by the repository operator.
— The physical control (security) of waste packages will be assured.
— Each waste package will be monitored for radiation.
— Radionuclide content will be measured by gamma scanning of selected waste packages.

III–2.5. Treatment of non-conformities

Non-conformities are evaluated and treated on a case-by-case basis in collaboration with the waste generator, regulator and repository operator. If the risk of a given non-conformity to the safe operation of the repository is significant, this fact will be declared in the safety report. According to this safety report, approval (or not) of the non-conformity is given by the regulator and the repository operator.

REFERENCE TO ANNEX III

[III–1] PRAZSKA, M., "Enhancing Radioactive Waste Management Capabilities", presented at the Regional Workshop on Waste Acceptance Criteria Development and Use, 23–27 May 2016, Bucharest, 2016.

Annex IV

**WASTE ACCEPTANCE CRITERIA DEVELOPMENT
METHODOLOGY: THE APPROACH IN SPAIN**

In Spain, there are two different radioactive waste repositories at El Cabril Disposal Centre: the low and intermediate level waste (LILW) repository and the very low level waste (VLLW) repository. El Cabril Disposal Centre is designed for the disposal of disposal units (DUs), which are concrete modules containing a certain number of waste packages contained in a solid matrix) and not for disposing waste packages directly; therefore, the DUs are the objects that need to comply with the waste acceptance criteria (WAC) [IV–1, IV–2]. WACs for the waste packages are derived from the relevant DU WAC in such a way that the waste packages automatically meet the DU WAC.

The DU WAC established for VLLW, LLW and ILW packages contains the following main sections:

— Acceptance criteria for waste packages;
— Methodology for accepting waste packages;
— Methodologies for waste package quality control.

IV–1. STRUCTURE AND CONTENT OF WASTE ACCEPTANCE CRITERIA

IV–1.1. Structure and contents of waste acceptance criteria for low level waste and intermediate level waste disposal units

There are two types of LILW DU, depending on the specific activity involved:

— Level 1 (LLW): The total mass, excluding shielding materials, is considered in the specific activity calculation.
— Level 2 (ILW): Specific activity determined exclusively with those DU components that fulfil the confinement limits.

A degree of activity heterogeneity is allowed for each DU (i.e. between and within waste packages). WAC relating to mechanical properties are more stringent for Level 2 DU than for Level 1 DU. In addition to these requirements, Level 2 DU has to be confined, and this means that the DU has to be able to limit the rate of release of radionuclides to values below the established limits. Allowable thermal cycles are also established for Level 2 DU.

The following criteria are incorporated in the content of the acceptance criteria for DU:

— Non-radiological criteria: No explosives, no substances that promote leaching and hinder the stabilization of cements, mortar or concrete, no high exothermic substances, etc.
— General criteria applicable to the radioactive waste package: Identification, degree of filling, acceptable containers, no free liquid, conditioned waste, dose rate limits, transport requirements, etc.
— Date of package production.
— Activity levels of packages.
— Nature of waste and waste form:
  • Ion exchange resins, evaporator concentrates, sludge solidified with cement.
  • Cartridge filters, ashes, dry sludge, immobilized with a mortar wall.

- Heterogeneous waste. Compactable waste will be pre-compacted inside the drum. Non-compactable waste will be conditioned by gap filling by means of grouting or using other substances with the objective of achieving compactness.

— Quality criteria.

As a function of the waste form, level of activity and age of the package, different quality criteria will be required, such as strength requirements, diffusion and leaching. Figures IV–1 and IV–2 show the quality criteria for LILW and VLLW, respectively.

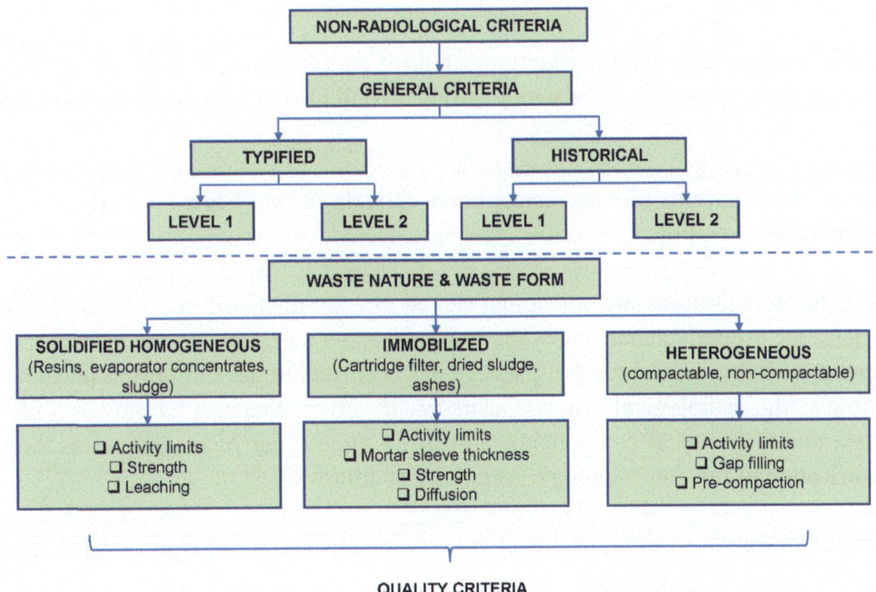

FIG. IV–1. Illustration of quality criteria for low and intermediate level waste as a function of the waste form, level of activity and age of package.

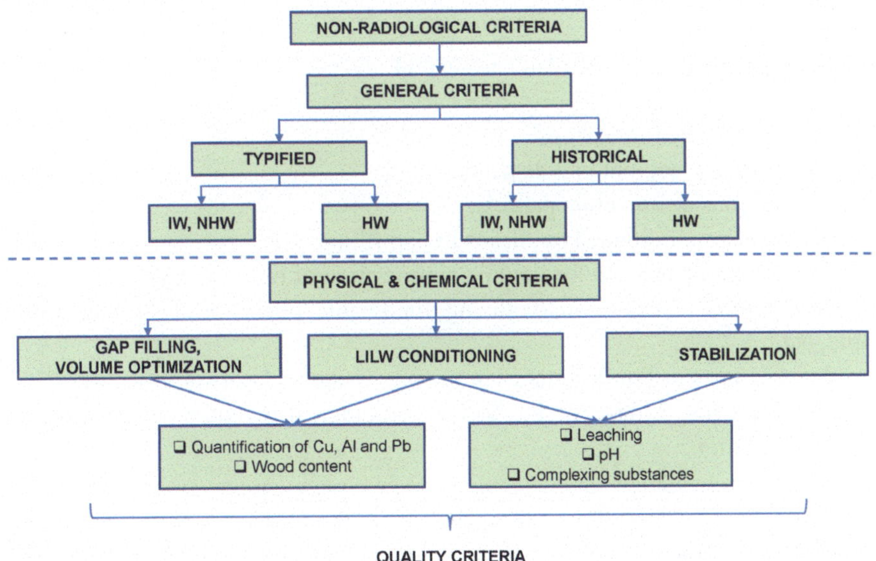

FIG. IV–2. Illustration of quality criteria for very low level waste as a function of waste form, activity level and package age. HW — hazardous waste; IW — inert waste; LILW — low and intermediate level waste; NHW — non-hazardous waste.

Level 1 waste package specific requirements are:

— Solidified homogeneous waste (resins, evaporated concentrates, sludge): Mechanical limits (compression, before and after immersion);
— Blocked waste (cartridge filters, dried sludge, ashes): Thickness of the mortar/concrete sleeve; mechanical limits (compression) of the sleeve;
— Heterogeneous waste: Compactable waste – segregation process; non-compactable waste – gap filling.

Level 2 waste package specific requirements:

— The specific requirements (quality requirements) of Level 2 waste packages are similar to those of Level 1 waste packages but generally more stringent and also include leaching limits and diffusion limits.

IV–1.2. Structure and contents of waste acceptance criteria for very low level waste disposal units

VLLW DU is, by definition, any object that can be directly disposed in a VLLW repository. In this case, a VLLW package properly generated by the producer could meet the DU acceptance criteria and, for this reason, can be considered DU once that package has been sent to the El Cabril repository.

In addition to the radiological risk associated with any radioactive waste, in VLLW the non-radiological risk, due to chemical components, are of the same order of magnitude as the radiological risks. The classification of the non-radiologic features is as follows:

— Inert waste (IW) shows no significant changes with time, such as minerals or substances in relation to natural soil substrate (soil, debris, etc.).
— Equivalent to non-hazardous waste (NHW) shows slow changes with time and slow environmental release rates, such as scrap iron and non-ferrous materials (pipes, equipment, textiles, plastics, etc.).
— Hazardous waste (HW) contains chemical contaminants that produce toxic risks in addition to radioactivity.

As in the case of LILW, this VLLW also has general criteria:

— VLLW DU will be accepted in batches.
— The container has to prevent material dispersion.
— HW requires a previous stabilization process.
— IW and NHW could be conditioned by means of gap filling.
— Non-pulverized IW could be used for volume optimization of other VLLW.
— No waste that is explosive, corrosive, oxidizing, flammable, pyrophoric, gas generating, putrescent, fermentable, etc.
— No encapsulated sources or liquid aqueous or organic waste without solidification.
— Each waste producer has to describe the VLLW radiological inventory of its waste justifying the method of determination.

The main radiological criterion is associated with dose criteria linked with the repository properties. If additional treatment is required at El Cabril repository, then this means that the treatment carried out by the producer did not meet the AC of the VLLW DU, and these packages cannot be considered compliant with the DU requirements when received at the El Cabril repository. Fig. IV–3 illustrates this concept.

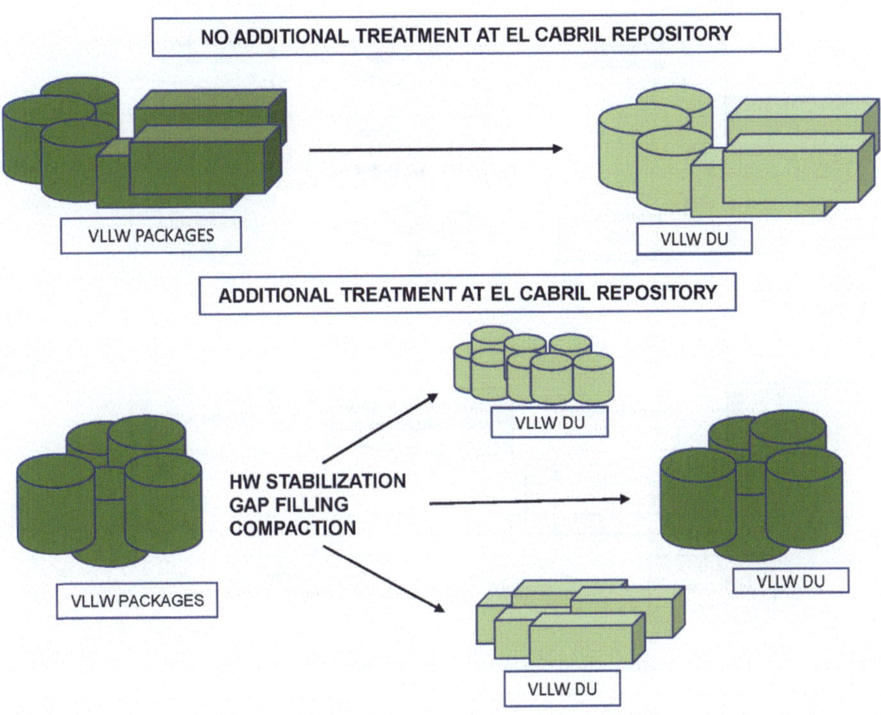

FIG. IV–3.   Illustration of the acceptance process for very low level waste (VLLW). DU — disposal units; HW — hazardous waste.

An additional concept arises in VLLW: the batch activity criterion. This means that not only do the activity limits for every package need to be fulfilled, but also that the batch of packages taken as a whole has to meet the additional activity criterion.

The following criteria have been defined for handling batches, in order to avoid inefficient processes:

— Packages containing hazardous waste and packages containing NHW are not allowed in the same batch.
— Waste packages requiring stabilization and those not requiring stabilization are not allowed in the same batch.

IV–2.   METHODOLOGY FOR ACCEPTING WASTE PACKAGES

Figure IV–4 illustrates the methodology for accepting both types of waste packages together with their interactions as described in this section.

IV–2.1. Application methodology for low and intermediate level waste acceptance

In the first stage of acceptance, the producer sends a waste package specification to Enresa for evaluation; this is known as the producer book (PB). It describes the basic features of the waste involved, such as:

— The nature of the waste;
— The way it was conditioned;
— The activity content and its evaluation;
— Quality assurance information.

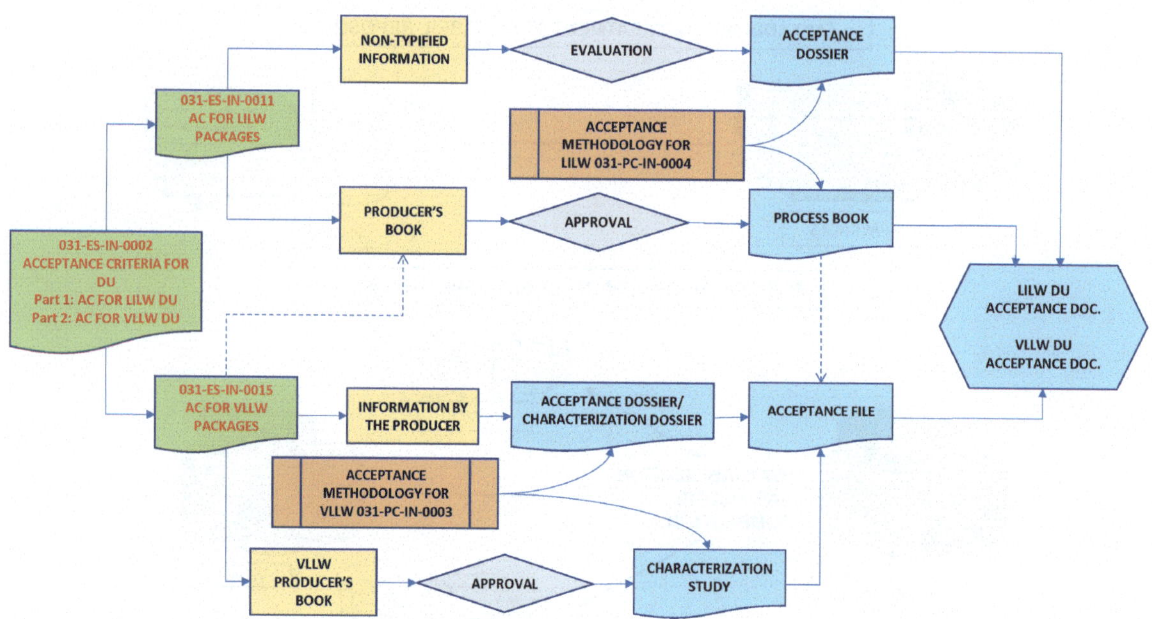

FIG. IV–4. Illustration of the waste acceptance criteria process flow for very low level waste (VLLW) and low and intermediate level waste (LILW). AC — acceptance criteria; DOC. — documentation; DU — disposal units.

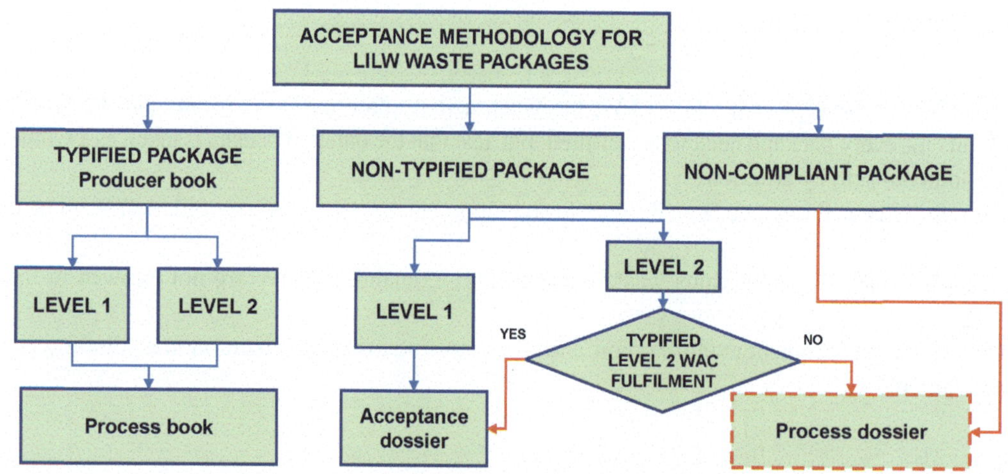

FIG. IV–5. Illustration of the acceptance process flow for low and intermediate level waste (LILW). WAC — waste acceptance criteria.

If the PB is approved by Enresa, the type of waste package involved is termed a *typified package*. In the case of historical packages, the producer sends a report to Enresa describing similar features regarding the production of those packages in the past. This type of historical package is termed a *non-typified package*. Both kinds of packages defined can also be Level 1 or Level 2 depending on the activity content.

In the case of typified packages, Enresa then evaluates the PB, checking whether the process described meets the package WAC or not. If so, Enresa generates the process book, containing the following subjects:

— Mechanical tests performed;
— Diffusion/leaching tests performed;
— Activity methodology description and evaluation;
— Transport requirements.

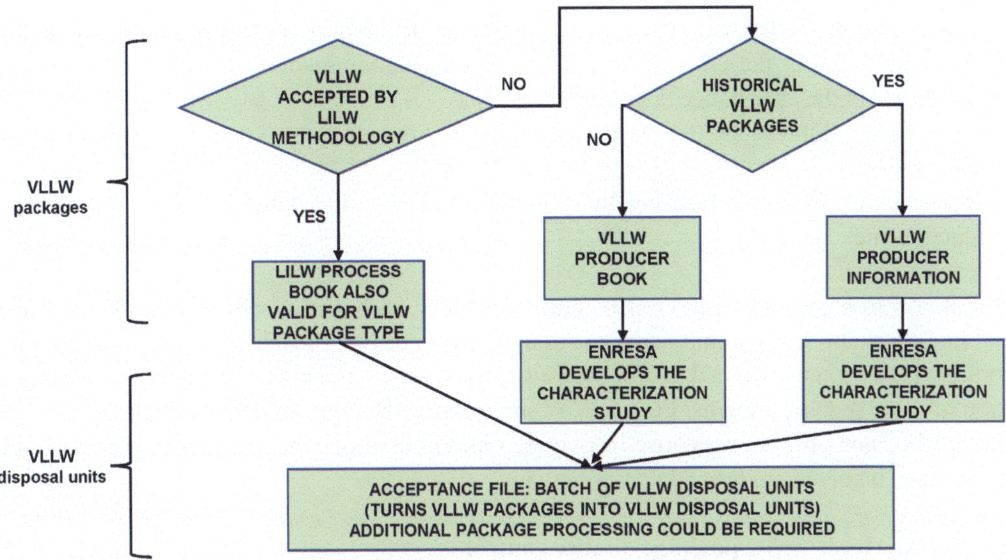

FIG. IV–6. Illustration of options for waste acceptance criteria implementation of VLLW packages. LILW — low and intermediate level waste; VLLW — very low level waste.

In the case of historical packages, Enresa produces the acceptance dossier. After checking the data sheets of produced packages from the waste stream involved, Enresa grants the acceptance of those packages in a process called document acceptance.

Finally, Enresa accepts the packages for transport to and disposal at El Cabril repository. This means that there is a transfer of responsibility from the producer to Enresa. Figure IV–5 shows the flow of the process involved in the methodology of acceptance.

IV–2.2. Application methodology for very low level waste acceptance

Acceptance of VLLW packages follows a process analogous to that of LILW. In addition to the concepts previously described, when some packages are produced to meet the framework of LILW criteria, it sometimes turns out that they can qualify as LILW packages. In this case, they do not need extra documentation for being accepted in the VLLW framework, because the LILW criteria are more restrictive than those for VLLW. Only the acceptance file is necessary for inclusion of these packages in a batch of VLLW packages. Figure IV–6 summarizes this process.

IV–3. METHODOLOGY FOR WASTE PACKAGE QUALITY CONTROL

IV–3.1. Application methodology for low and intermediate level waste packages

Enresa implements two kinds of quality controls in relation to waste package acceptance:

— In situ controls carried out in the nuclear power plant;
— Off-site controls carried out at El Cabril laboratory.

There are four broad types of production controls (in situ controls):

(1) Resources control: Carried out when a new package type is launched by a producer, to verify if the producer resources are in agreement with the PB;

(2) Activity control: Performed when there are large differences between producers and Enresa calculations to verify the effectiveness of the activity determination methodology approved for a package type in the acceptance documents;

(3) Process audit: Represents the verification of the overall waste package production process documentation;

(4) Process control: Performed during the production of one or a batch of waste packages, at the production site.

Off-site controls performed at the El Cabril repository, technical verification tests, are carried out on actual packages and test gamma activity, strength properties and leaching values. The frequency of these controls is previously planned in order to cover every waste stream per producer as a function of the generation of these specific streams. Figure IV–7 summarizes the types of off-site controls.

When WAC non-conformities arise from these types of controls, the acceptance process is rejected, and the packages involved are termed non-compliant.

IV–3.2. Very low lever waste package quality control

In a similar methodology used for LILW, Enresa carries out in situ and off-site controls for the VLLW waste. There are three types of in situ controls:

(1) Audit control: For a generator with VLLW volumes above 200 m^3/year, consists of document verification of the overall process of waste package production;

(2) Activity control: For every 500 m^3 or every five years, performed on a data sheet provided by the producer prior to the acceptance process;

(3) Process control: At the production site, during the production of one or a batch of waste packages, for quantities above 100 m^3/year.

<table>
<tr><td align="center">TESTS</td></tr>
<tr><td>

❏ Visual and photographic inspection of package
❏ Dose rate at contact
❏ Container weight
❏ Dimensional control (*)
❏ Verify values of ^{60}Co and ^{137}Cs declared by the producer (whole package gamma spectrometry)
❏ Visual X ray inspection (**)

</td></tr>
<tr><td>

❏ Determination of weak β emitters
❏ Determination of α emitters
❏ Chemical determination

</td></tr>
<tr><td>

❏ Absence of free liquid
❏ Correct setting of the matrix
❏ Uniaxial mechanical resistance (***)

</td></tr>
</table>

FIG. IV–7. Illustration of verification tests performed for different types of wastes. (*) — Only for non-irradiating waste packages; (**) — If needed; (***) — Additional checks may also be considered.

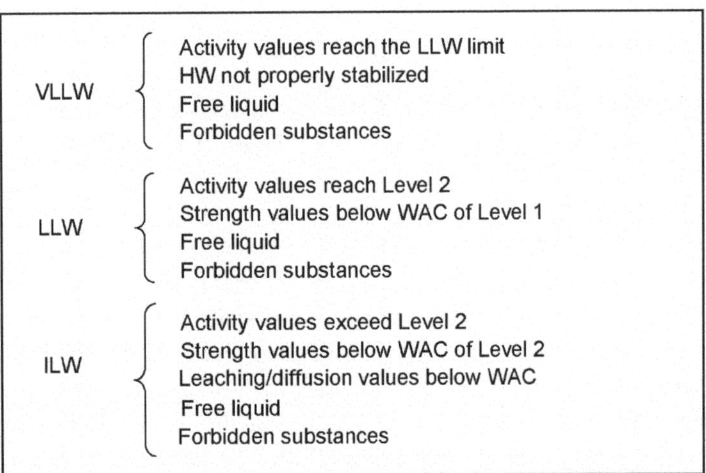

FIG. IV–8. Features that can be the subject of a non-conformity. HW — hazardous waste; ILW — intermediate level waste; LLW — low level waste; VLLW — very low level waste; WAC — waste acceptance criteria.

Off-site controls (checking tests) performed at El Cabril repository are carried out on waste packages; the tests include gamma spectrometry, radiochemical analysis and leaching tests.

IV–3.3. Treatment of non-conformities

Non-conformities can be detected in the following cases:

— During the waste package acceptance process;
— During the quality control acceptance process.

Some features that can be the subject of non-conformities are shown in Fig. IV–8. Where non-conformities occur, an additional non-conformity remediation programme will be developed to prevent future failures during the acceptance processes.

IV–4.   FUTURE ISSUES

Public opposition is the main problem for siting new storage or disposal facilities. Site selection activities for a deep geological repository have been suspended until definitive management methods and corresponding regulatory process are established and a better social acceptance can be reached. In the near future, a significant effort will be made to develop new safety guidelines or to endorse IAEA Safety Guides from the RADWASS Programme. Effort will also be made to develop a national regulatory framework for spent fuel and HLW management.

REFERENCES TO ANNEX IV

[IV–1]   OECD NUCLEAR ENERGY AGENCY, The Regulatory Control of Radioactive Waste Management, REV2, OECD NEA, Paris (2003).
[IV–2]   NAVARRO M., LEGANÉS J.L., "Waste Acceptance Criteria. Spanish approach. L&ILW and VLLW", IAEA TC Regional Workshop, Vilnius, 2009.

Annex V

**WASTE ACCEPTANCE CRITERIA DEVELOPMENT
METHODOLOGY: THE APPROACH IN JAPAN**

V–1. STRUCTURE AND CONTENT OF WASTE ACCEPTANCE CRITERIA

The main disposal centre in Japan has been operational since 1992 and is part of the Japan Nuclear Fuel Limited (JNFL) nuclear fuel cycle facilities. After the prescribed period of on-site storage on the nuclear power plant's premises, the LLW waste (consisting mainly liquid waste concentrates, spent resins, waste sludge and other operational wastes placed in 200 L metal drums) are shipped to the disposal facility.

For other types of wastes, such as transuranic (TRU) waste or those generated in the uranium fabrication facilities, the disposal related safety regulations were established in 2000 and amended in 2007 to include these waste types [V–1].

In Japan, LLW typically refers to all radioactive waste other than high level waste (HLW) (which includes liquid and vitrified waste from spent fuel reprocessing). The Japanese Nuclear Safety Commission (JNSC) has issued a document that provides for upper bounds (considering the latest knowledge in the international community and keeping the public exposure well within the reference values) of concentration of radioactive elements in waste packages. Radioactive waste is categorized as indicated in Table V–1, and the inventory is referenced in Ref. [V–2].

WAC for near surface disposal can be divided into two types: technical criteria and quality management criteria. The technical criteria were independently developed according to the combination of the type of waste package and the type of disposal facility. The quality management criteria are based on JEAC 4111 (Japan Electric Association Code 4111: Quality assurance codes for safety in nuclear power plants) [V–3], ISO9001 [V–4] and IAEA Safety Standards Series No. GSR Part 2 [V–5].

Generic requirements for waste packages, which do not depend on the type of disposal facility for LLW, have been stipulated by law as performance based technical criteria for waste packages.

Additional requirements from the perspective of the design and safety assessment of each individual disposal facility (near surface disposal and sub-surface disposal) are added as acceptance criteria for disposal.

For WAC compliance, the waste packages will have to fulfil the generic requirements, which form the basis of the WAC, the additional requirements (the basis for disposal) and transportation regulations. In Japan, however, where special containers are exclusively used for transportation, most of the transportation requirements are imposed on those containers and not on the waste packages.

TABLE V–1. CATEGORY CLASSIFICATION OF WASTE

Category of wastes	Cumulative amount of wastes as of 2017
High level waste (vitrified waste)	6 206 canisters (vitrified waste)
Waste generated from nuclear reactors:	
LLW containing comparatively high radioactivity (core internal structure, etc.)	Control rods (10 951) Channel boxes, etc. (71 999)
LLW	708 883 drums (200 L) at nuclear power plants (290 000 drums were disposed at Rokkasho Mura Disposal Facility)
Very low level waste	1 670 t were disposed at JAERI's Tokai site
Uranium waste	52 810 drums (200 L)
Waste originated from medical, industrial and research facilities	511 012 drums (200 L)*

* This number includes TRU waste and Japan Nuclear Cycle Development Institute waste originating from uranium fabrication facilities.

V–2. PROCESS OF ACCEPTANCE

Waste generators apply for disposal to the relevant authorities. It is stipulated by law that the authorities will verify the application by inspection. Waste generators supply documents describing the quality of solidification material, quality of container, waste conditioning method, test result of burial, load resistant strength of waste package and activity determination method to the authorities after inspection.

In Japan, the conformity of waste packages to the WAC is verified by governmental authority in parallel with waste package preparation management and shipping inspections carried out by waste generators. Practically, this verification is entrusted to third-party agencies by the governmental authority.

V–3. CONTROLS

Appropriate quality control is carried out according to the type of waste and the method of conditioning. Cement solidification of concentrates is used here as an example to introduce the method and system for verifying the adaptability to WAC in Japan.

V–3.1. Process control of waste conditioning in the operational phase

At the solidification operational phase, the following analysis and operation data are obtained and recorded:

— Analysis of liquid waste: Composition of nuclides in liquid waste, concentrations of principal elements in liquid waste;
— Operational record: Waste-to-binder mixing ratio, mixing speed, etc.

V–3.2. Waste generator inspection in the predisposal phase

In disposal of solidified waste packages, waste generators inspect all waste packages and record the activity concentrations of ^{60}Co and ^{137}Cs that are considered the key radionuclides, surface dose rate, compressive strength, surface contamination, external appearance and package weight.

Waste generators then apply for disposal to the relevant authorities. The law stipulates that the authorities will verify the application by inspection. An automatic inspection facility has been installed in each nuclear power plant to accomplish this purpose [V–6].

V–3.3. Compliance verification

In considering waste generators' applications for permission for disposal of waste packages, the authorities concerned independently verify the conformity of waste packages to the WAC by checking records of testing, assessment, waste package manufacturing control and inspections conducted by waste generators in accordance with the flow chart shown in Fig. V–1.

Waste generators carry out waste package manufacturing controls and perform waste package inspections to ensure conformity with the WAC. The results of the waste package manufacturing control and inspections are submitted to the concerned authorities in order to obtain permission for disposal of the waste packages.

V–3.3.1. Governmental verification in the pre-operational phase

The concerned authorities independently verify the results of the waste package manufacturing control and inspections that were performed by the waste generators. The authorities assess reports of

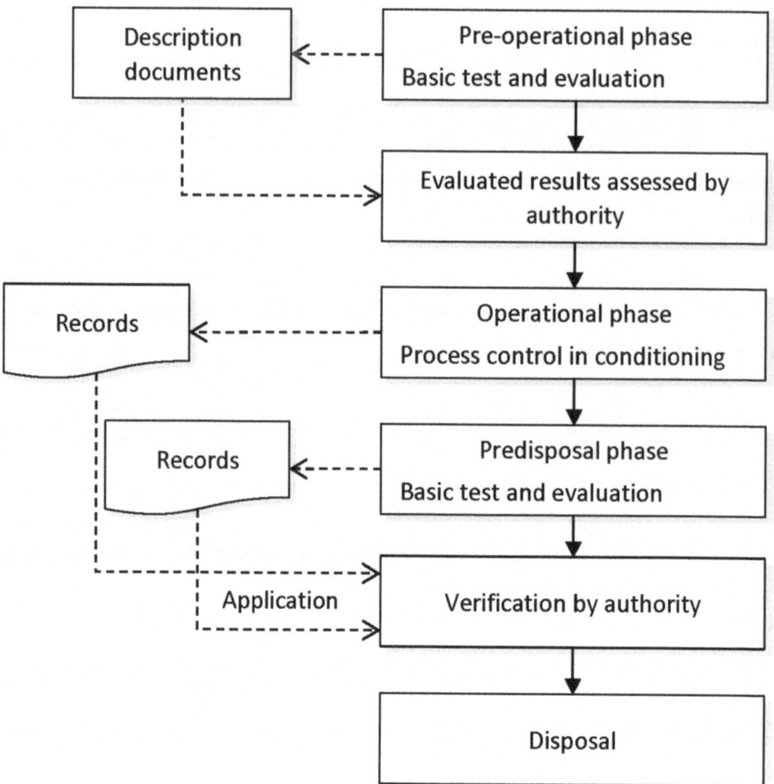

FIG. V–1. Basic flow chart of controlling and verifying waste acceptance criteria.

the results concerning the waste manufacturing method, properties of the waste form, strength of waste packages, activity determination method and radioactive data.

V–3.3.2. *Governmental verification in the predisposal phase*

The authorities concerned independently verify the results of the waste package manufacturing control and waste generators' inspections by applying the following methods:

— 100% inspection: Checks on inspection and operation records, visual inspection of waste packages;
— Sampling inspection: Inspection of activity concentration, surface dose rate, compressive strength, surface contamination and weight;
— Destructive inspection: Analysis of the concentration of nuclides by sampling analysis.

V–3.4. Treatment of departures and non-conformities

The waste generator will treat departures from the design condition of the waste management facility, such as the change of waste property, according to Ref. [V–3], and will re-evaluate and review the design condition from a viewpoint of influencing the waste processing.

The waste generator has the responsibility to ensure the conformity of the waste packages to the WAC. The waste generator will treat the measures for the non-conformity to the WAC according to the procedure that is required by Ref. [V–3].

V–4. FUTURE ISSUES

There are currently plans to construct a sub-surface disposal facility, at around a 50 to 100 m depth, for ILW (referred to as 'relatively higher level LLW').

REFERENCES TO ANNEX V

[V–1]   Nuclear Waste Management Organization of Japan (NUMO), The Act on Final Disposal of Specified Radioactive Waste" (Final Disposal Act), 2000, amended in 2007 to include TRU.

[V–2]   GOVERNMENT OF JAPAN, National Report of JAPAN for the Sixth Review Meeting, IAEA, Vienna (2017), https://www.iaea.org/sites/default/files/national_report_of_japan_for_the_6th_review_meeting_-_english.pdf

[V–3]   JAPAN ELECTRIC ASSOCIATION, Rules of Quality Assurance for Safety of Nuclear Power Plants, JEAC 4111-2003, JEA, Tokyo (2003).

[V–4]   INTERNATIONAL ORGANIZATION FOR STANDARDIZATION, Quality Management Systems — Requirements, ISO Standard No. 9001:2015, ISO, Geneva (2015).

[V–5]   INTERNATIONAL ATOMIC ENERGY AGENCY, Leadership and Management for Safety, IAEA Safety Standards Series No. GSR Part 2, IAEA, Vienna (2016), https://doi.org/10.61092/iaea.cq1k-j5z3

[V–6]   FINSTER, M., KAMBOJ, S., International Low-Level Waste Disposal Practices and Facilities, ANL-FCT-324, Argonne National Laboratory, Lemont, IL (2011), https://doi.org/10.2172/1033484

Annex VI

**WASTE ACCEPTANCE CRITERIA DEVELOPMENT
METHODOLOGY: THE APPROACH IN HUNGARY**

VI–1. INTRODUCTION TO THE NATIONAL RADIOACTIVE WASTE REPOSITORY

The Public Limited Company for Radioactive Waste Management (PURAM) is the Hungarian radioactive waste management organization, responsible for the safe collection, handling, storage and disposal of all kinds of radioactive wastes in the country. PURAM commenced operating at the National Radioactive Waste Repository (NRWR) for low and intermediate level waste (L/ILW) at Bátaapáti in 2012.

The original purpose of the facility was to receive the L/ILW generated in Paks Nuclear Power Plant, but recently a need for disposal of institutional wastes has emerged because the capacity of the existing facility destined for institutional wastes has run short.

The national repository is sited in granite host rock at a depth of 250 m. In 2020, four emplacement chambers (I-K1, I-K2, I-K3, I-K4) were already excavated, and at least two more chambers (I-N1, I-N2) are planned to be excavated. I-K1 has already been loaded with radioactive waste and the continuation of emplacement in I-K2 was foreseen in 2023.

The safety concept of the facility has evolved continuously; thus, the safety concept differs between chambers:

— I-K1: The multibarrier system involves the waste matrix, the waste package (the primary package is a carbon steel drum), reinforced concrete over-container, buffer material, backfill material and the geological formation itself.
— I-K2, I-K3, I-K4, I-N1, I-N2: The multibarrier system involves the waste matrix, the primary waste package (compact waste package (CWP)), the reinforced concrete vault in the chamber, backfill material and the geological formation itself.

The main difference between the safety concepts of I-K1 and the other chambers is the role of reinforced concrete structures. In I-K1, the waste package itself was made of reinforced concrete; in the other chambers the safety function is provided by the reinforced concrete walls (and ceiling) of the vault. This conceptual change enhanced the efficiency of the emplacement system (the radioactive/inactive material balance in a chamber) without jeopardizing the overall long term safety of the system. The difference between the concepts is shown in Fig. VI–1.

FIG. VI–1. Reinforced concrete containers in the I-K1 chamber, and the reinforced concrete walls of other chambers in the Hungarian National Radioactive Waste Repository (NRWR). Courtesy of the Public Limited Company for Radioactive Waste Management (PURAM), Hungary.

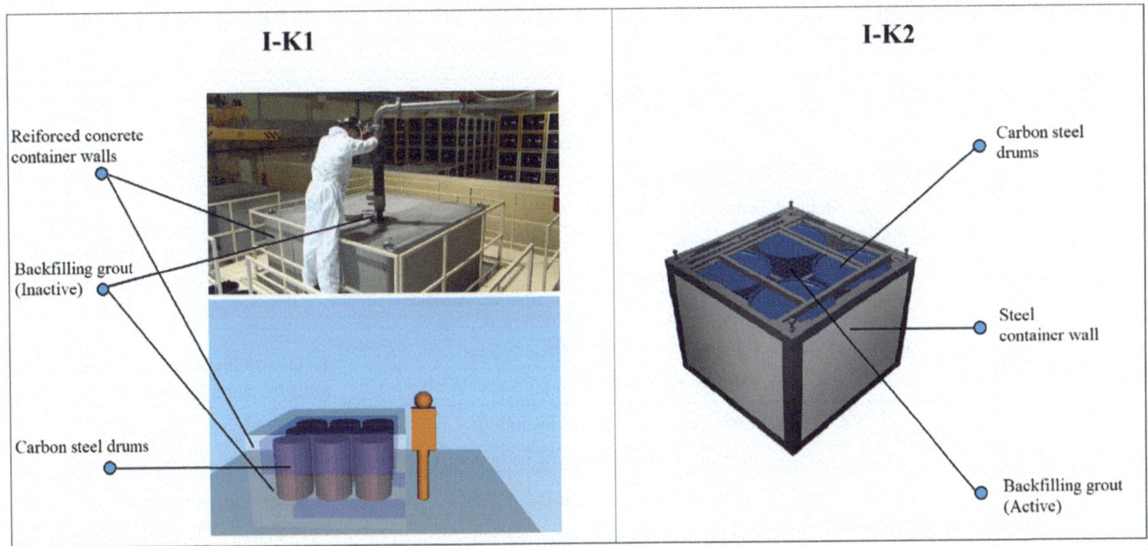

FIG. VI–2. Waste package types and differences. Courtesy of the Public Limited Company for Radioactive Waste Management (PURAM), Hungary.

The CWP is a small sized (134 cm × 134 cm × 103.2 cm) steel container, which contains four pieces of 200 L carbon steel drum. The backfilling material (grout) in the package is made using liquid radioactive waste. The differences between the waste package types are shown in Fig. VI–2. Non-containerized 200 L drums are also allowed to be disposed of in chamber I-K2.

VI–2. WASTE ACCEPTANCE CRITERIA IN THE HUNGARIAN NATIONAL RADIOACTIVE WASTE REPOSITORY

According to Hungarian regulations, WAC is issued as part of the safety case. Key elements of the WAC for chambers I-K1 and I-K2 are summarized in Table VI–1.

VI–3. WASTE ACCEPTANCE CRITERIA DEVELOPMENT METHODOLOGY

VI–3.1. Evolution of waste acceptance criteria

The development of the WAC is an iterative process, carried out in conjunction with the facility design and safety assessments and with close cooperation between waste generator and repository operator:

— A generic WAC was developed at the conceptual design stage of the facility.
— The first site specific draft of the NRWR's WAC was issued in 2005 with several parameters based on international recommendations and international best practices and activity concentration limits based on the safety assessment.
— The WAC was upgraded through the preliminary and detailed design and construction stages at each stage in the licensing process and in line with each revision of the associated safety case.
— The operating WAC (introduced in Section VI–2) is derived from the actual safety case for the operational licence [VI–1].

TABLE VI–1. WAC FOR THE I-K1 AND I-K2 CHAMBERS OF THE HUNGARIAN NATIONAL RADIOACTIVE WASTE REPOSITORY [VI–1]

Properties (of)	Requirements		Goal
	Type	Criteria	
Packaging	The type of the package depends on the place of disposal	I-K1: 200 L drums in reinforced concrete containers I-K2: CWPs I-K2: 200 L drums	To ensure the safe transport and handling of the waste packages
Waste form	Gas production	Limitations regarding to organic and biodegradable materials Max. overpressure in a package: 0.5 bar	To ensure the integrity of the multibarrier system of the repository.
	Gas content	Compressed gases are excluded Special treatment (use of sorbents) if needed	
	Heat production	HLW (2 kW/m^3) excluded Additional assessments over 20 W/m^3	To ensure the integrity of the waste form and the multibarrier system of the repository.
	Dust content	Max. 1 m/m%	Operational safety and reducing the releases to the environment
	Cavity volume	Max. 10 V/V %	Reducing the releases to the environment
	Free liquids	Exclusion of free liquids Special cases: Max. 1 V/V%	
	Corrosive materials	Max. 1 m/m%	
	Chelating and complexion agents	Max. 1 m/m%	Operational safety and reducing the releases to the environment
	Hazardous materials	Excluded (HP1, HP2, HP3, HP8, HP9)[a]	
	Combustibility	Limited amounts Special conditioning techniques if needed	
	Compressive strength	10–30 N/mm^2 if cement based condition technique is applied	
	Chemical stability	Special conditioning techniques, if necessary	Reducing the releases to the environment
	Leaching	$D_{eff} < 1E - 7$ cm^2/s for conditioned, stabilized waste forms	
	Homogeneity	Need to be achieved	Operational safety
Radiological	Isotope composition	Activities for each isotope of the waste package need to be known for a reference date	Operational (normal operation and design basis accidents) and long term safety
	Isotope content	Activity concentration limits	Operational (normal operation and design basis accidents) safety
	Fissionable material[b] content	Activity concentration limits for fissionable isotopes	Exclusion of criticality
	Surface dose rate	10 mGy/h > dD/dt	Operational safety
	Surface contamination	Max. 4 Bq/cm^2 for β and γ Max. 0.4 Bq/cm^2 for α	Operational safety and reducing the releases to the environment

[a] Hazard Property Codes: HP1: Explosive; HP2: Oxidizing; HP3: Flammable; HP8: Corrosive; HP9: Infectious.
[b] Fissionable materials include all nuclides that are capable of induced fission, including fissile materials.

TABLE VI–2. BASIS OF WASTE ACCEPTANCE CRITERIA FOR THE I-K1 AND I-K2 CHAMBERS OF THE HUNGARIAN NRWR

Properties (of)	Requirements	Source
Packaging	Package descriptions	Operational experience, International best practice
Waste form	Gas production	International best practice
	Gas content	International best practice
	Heat production	Legal requirement
	Dust content	International best practice
	Cavity volume	International best practice
	Free liquids	International best practice
	Corrosive materials	International best practice
	Chelating and complexion agents	International best practice
	Hazardous materials	International best practice
	Combustibility	International best practice
	Compressive strength	International best practice
	Chemical stability	International best practice
	Leaching	International best practice + safety assessment
	Homogeneity	International best practice
Radiological	Isotope composition	Legal requirement
	Isotope content	Safety assessment
	Fissionable material content	Safety assessment
	Surface dose rate	Legal requirement
	Surface contamination	Legal requirement

VI–3.2. Basis of waste acceptance criteria derivation

WAC can be characterized according to the types of sources being used to derive them:

— WAC based on legal requirements (undesirable substances, free liquids, burnable and gas generating materials, etc.);
— WAC connected with operational experiences or based on international good practice (compression strength of matrix, package weight, content of complexing agents, container parameters, etc.);
— WAC directly derived from safety assessments (requirements regarding to activity concentrations, activity limits, etc.).

The sources of the Hungarian NRWR's WAC are shown in Table VI–2.

VI–3.3. Waste acceptance criteria based on safety assessment

Derivation of WAC and waste emplacement requirements (WER) is essential for the overall safety of a radioactive waste repository. WAC and WER can be based on different safety assessments that were being made in correspondence with the repository:

— Safety assessments regarding to the design basis accidental scenarios;
— Long term (post-closure) safety assessments;
— Criticality safety assessments.

During the derivation of WAC and WER, measures had to be taken to identify and separate the time frames and the locations when and where a certain safety goal is applicable to define the criteria which guarantees the fulfilment of the safety goals.

The WAC are activity concentrations interpreted for each isotope of each waste stream (independent 200 L drums, radioactive grout of CWPs, 200 L drums in CWPs). The application of WAC ensures that the relevant regulatory limits (dose constraints both for employees and representatives of the public) will not be exceeded even during design basis accidents.

The WER are activity limits derived from the long term and criticality safety analyses only for the safety relevant isotopes of each waste stream. Activity limits can be interpreted for an emplacement chamber or for a vault section in the chamber, depending on the safety goal to be achieved.

Due to the nature of the different safety goals, the definition of different and independent safety criteria is necessary.

Safety criteria regarding WAC are inspected during the handover of waste packages, while safety criteria concerning WER are monitored during the operation of the disposal facility.

In order to ensure the continuous operation of the disposal facility, the risk of accepting non-disposable waste packages is minimized by linking WAC and WER.

VI–3.3.1. Waste acceptance criteria derivation methodology

Radionuclide activity related WAC were derived from design basis accidental scenarios, using the same methodology for each isotope and scenario.

Scenario specific dose conversion factors were determined as the ratio of the dose contribution and the activity of the given isotope in the given scenario. Dividing the necessary dose constraint (depending on the scenario) with the dose conversion factor gives the maximum allowed activity in a waste package for each isotope in the given scenario. Activity concentrations can be determined using the maximum allowed activities and the volume of a waste package.

Different accident scenarios provide different activity concentrations. The activity concentration accepted as WAC would be the minimum value of the range in order to ensure that the WAC considers the impacts of all accidental scenarios.

Figure VI–3 summarizes the radionuclide activity related WAC derivation methodology for a given waste stream.

The application of WAC ensures that the dose contribution of each isotope will be under the dose constraint during every possible accident scenario. However, it is not sufficient if more than one radioactive isotope is present in the waste package.

Fulfilling the criteria of Eq. (VI–1) guarantees that the total radiological impact of the mixture of isotopes will not exceed the dose constraint:

$$\forall j : \sum_i \frac{Ac_{i,j}}{AC_{i,j}} < 1 \tag{VI–1}$$

where
$Ac_{i,j}$ is the activity concentration of isotope i in waste stream j (Bq/m^3);
and $AC_{i,j}$ is the WAC of isotope i in waste stream j (Bq/m^3).

The following design basis accidents are taken into account during the WAC development process:

— S1: Drop of CWP during waste manipulation;
— S2: Drop of 200 L drums (four drums at a time);
— S3: Ignition of non-containerized drums (burning of ten drums at a time);
— S4: Accidental atmospheric release of radionuclides due to S1–S3 scenarios;
— S5: Accidental liquid discharge of radionuclides due to the S1 or S2 scenario.

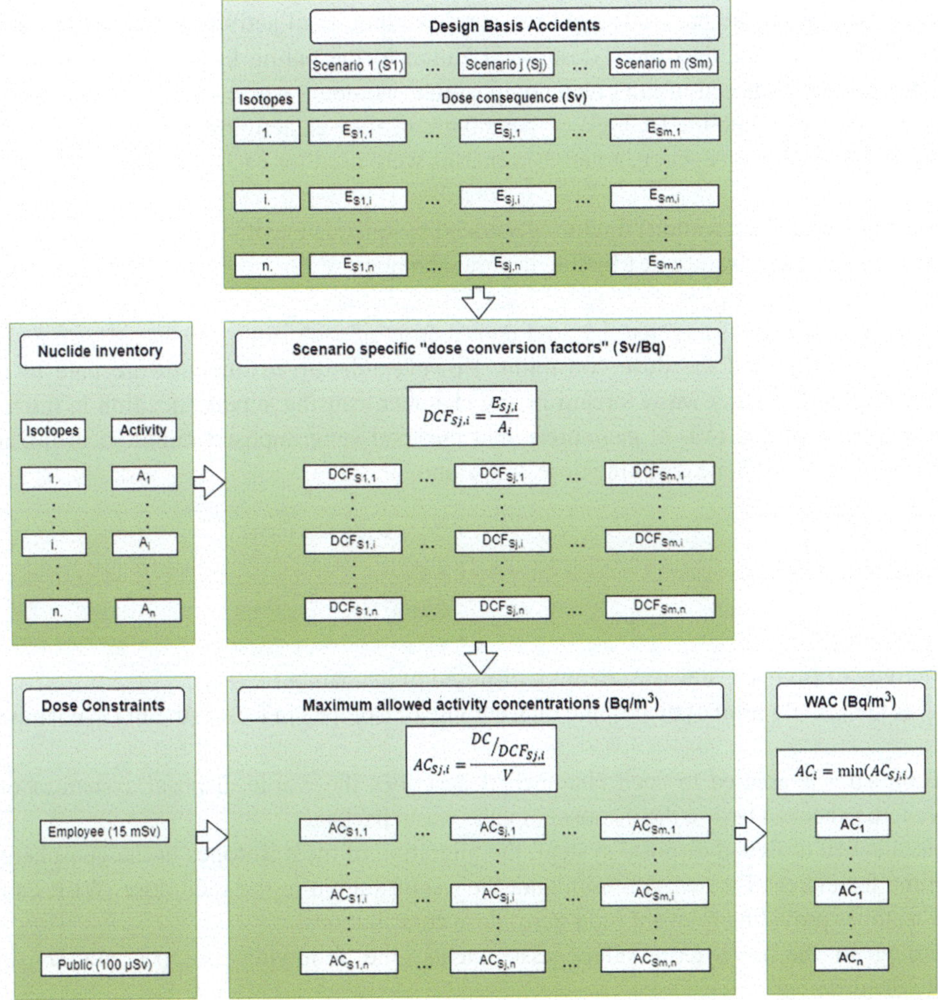

FIG. VI–3. Waste acceptance criteria (WAC) derivation methodology for a given waste stream. Courtesy of the Public Limited Company for Radioactive Waste Management (PURAM), Hungary. AC — activity concentration; DC — dose constraint; DCF — dose conversion factor; $E_{Sj,I}$ — dose consequence of isotope i during design basis accident j.

In order to ensure the continuous operation of the disposal facility, additional scenarios from WER were generated. As an arbitrary rule regarding the relevant isotopes, the maximum of 10% of each WER value is permitted in a waste package.

VI–3.3.2. Waste emplacement requirements

The waste emplacement requirements (WER) are activity limits, derived from long term safety assessments and criticality safety assessments. WER, unlike WAC, would not be applied to single waste packages. WER are interpreted in a chamber, or in a segment of a vault, depending on the type of assessment they are derived from.

(a) Waste emplacement requirements derived from long term safety

The peak dose of each isotope of each chamber's waste stream was identified based on the long term safety assessment models for a reference inventory. Special dose conversion factors (DCF) were determined as the ratio of the peak dose and the corresponding total activity emplaced in the chamber. This special DCF shows the maximum dose contribution to the public in the distant future (~10 000 years) of 1 Bq activity disposed in the given composition (isotope – waste stream – chamber). Dividing the public dose constraint with the DCF gives the maximum allowed activities.

During the process the most conservative scenarios were used for each chamber:

— Chamber crossing (intersecting) fracture generated by seismic event;
— Unknown conductive feature nearby the disposal chamber.

Application of WER ensures that the dose contribution of each isotope, waste stream and chamber composition will not exceed the dose constraint. However, dissolved nuclides are able to reach the biosphere originating from any waste stream in any chamber, causing superimposition in the total dose. Fulfilling the criteria of Eq. (VI–2) guarantees that the total superimposed radiological impact of the whole disposal system will not exceed the dose constraint:

$$\sum_{i,j,k} \frac{A_{i,j,k}}{ALIM_{i,j,k}} < 1 \qquad\qquad\qquad (VI–2)$$

where
$A_{i,j,k}$ is the activity of isotope i in waste stream j, disposed of in chamber k (Bq);
and $ALIM_{i,j,k}$ is the activity limit of disposable total activity of isotope i in waste stream j in chamber k (Bq).

Since the sum is applied to each chamber, it describes the whole disposal system at once. The criteria have to be checked before the disposal of each waste package.

Due to the well designed disposal system, the migration of most isotopes in the geosphere is slow, providing time for radioactive decay, minimizing their contribution in the total dose. WER can only be derived for isotopes providing relevant (non-zero) dose contributions.

According to the long term safety assessments, the following isotopes are considered to contribute to the dose:

— ^{14}C $(T_{1/2}$: 5 730 y);
— ^{36}Cl $(T_{1/2}$: 301 300 y);
— ^{40}K $(T_{1/2}$: 1 280 000 000 y);
— ^{41}Ca $(T_{1/2}$: 140 000 y);
— ^{79}Se $(T_{1/2}$: 65 000 y);
— ^{129}I $(T_{1/2}$: 15 700 000 y).

(b)   Waste emplacement requirements derived from criticality safety

Exclusion of criticality in the waste packages as well as in the disposal system is a fundamental safety criteria. Criticality in L/ILW packages is usually not a significant concern. However, local accumulation of fissionable materials during the lifetime of the disposal system is a complex issue. Moreover, it is a regulatory requirement to assess the criticality.

In order to address the uncertainty caused by the possible migration and accumulation of the fissionable materials after the closure of the disposal chambers, in order to be on the safe side, an extremely conservative methodology was developed.

A special set of fissionable isotopes has been determined, called 'WER amounts'. WER amounts are a set of fissionable isotopes, which can form a critical system.

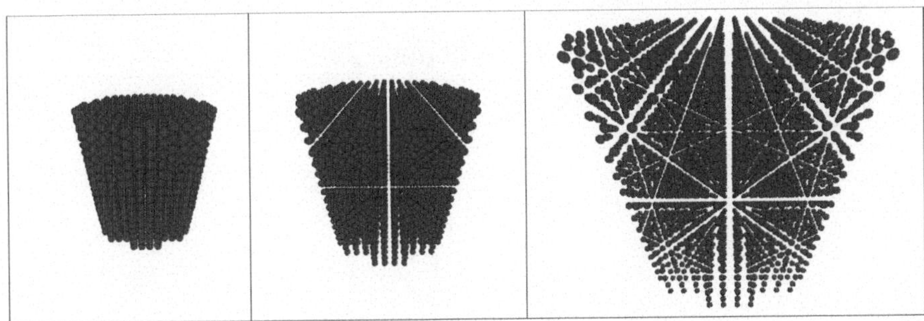

FIG. VI–4. 'Cell' geometry with different lattice pitch. Courtesy of the Public Limited Company for Radioactive Waste Management (PURAM), Hungary.

TABLE VI–3. WER AMOUNTS FOR FISSIONABLE ISOTOPES

Isotope	WER (Bq)	Isotope	WER (Bq)	Isotope	WER (Bq)
Th-229	8.00E+10	U-236	—[a]	Pu-241	1.80E+14
Th-230	—[a]	Pu-236	3.50E+14	Pu-244	5.00E+08
Pa-231	—[a]	Np-236	4.35E+10	Am-241	1.20E+14
Th-232	—[a]	Np-237	3.27E+09	Cm-242	1.00E+16
U-232	2.25E+13	U-238	—[a]	Am-243	7.00E+12
U-233	9.00E+09	Pu-238	7.00E+13	Cm-244	5.25E+14
U-234	2.00E+11	Pu-239	2.00E+12	Cm-248	1.00E+11
U-235	3.20E+08	Pu-240	2.30E+12	Cf-252	3.20E+14

[a] —: data not available.

The multiplication factor (k_{eff}) of the system was investigated under the following conditions:

— Two main type of the geometry was separately examined:
  - Block: The WER amounts are forming one sphere.
  - Cells: The WER amounts are separated in many sphere-shaped cells, forming a cylinder.
— The WER amounts contain fissionable materials without contaminations (pure heavy metal material composition).
— The fissionable materials are surrounded by water (also without contaminants).
— The correlation between the fissionable material/water ratio and the multiplication factor was examined by changing the pitch of the lattice of the cell mesh (Fig. VI–4).
— MCNP 6.1[1] was used in the calculations.

The WER amounts are shown in Table VI–3. Isotopes without defined WER amounts (e.g. ^{230}Th) are not considered a hazard for criticality safety.

The correlation between the multiplication factor and the pitch of the lattice, as well as the maximum k_{eff} of the cell-type geometry configuration of the WER amounts, is shown in Fig. VI–5. The maximum k_{eff} for block-type geometry is 0.79189 (SD: 0.00064).

---

[1] MCNP: Monte Carlo N-Particle Transport Code; https://mcnp.lanl.gov/

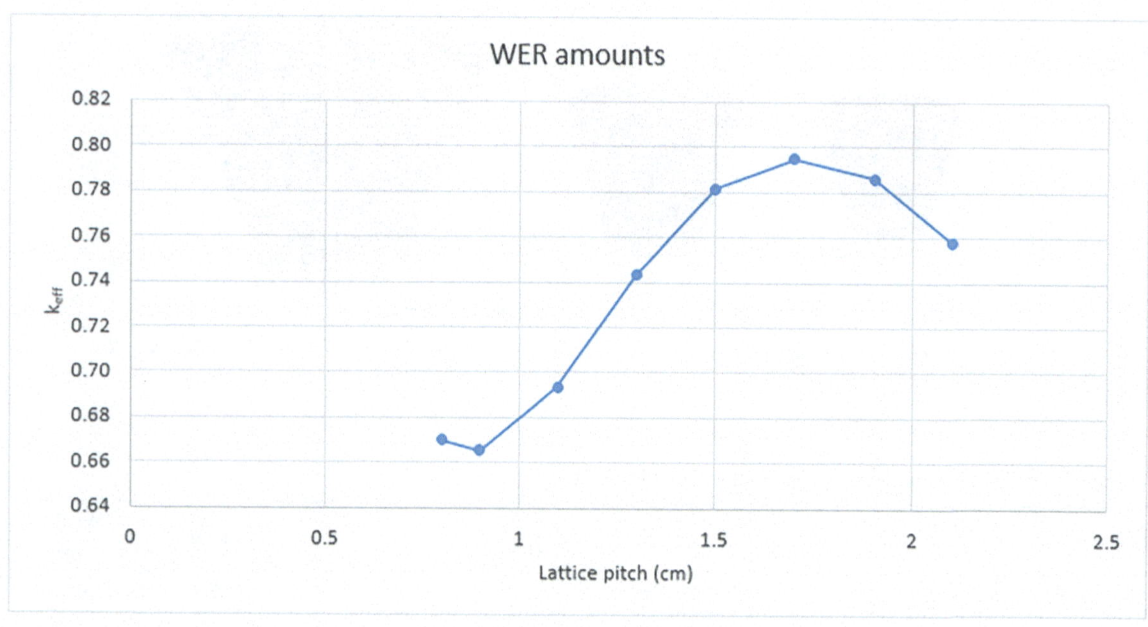

FIG. VI–5. Multiplication factor–lattice pitch correlation for waste emplacement requirements (WER) amounts. Courtesy of the Public Limited Company for Radioactive Waste Management (PURAM), Hungary.

WER amounts are the total activities allowed to be disposed of in well confined spaces, separated by engineered barriers such as reinforced vault walls. In the case of NRWR, the vault segments of the chambers are considered such spaces.

The fulfilment of the criteria shown in Eq. (VI–3) guarantees that every possible formation of locally accumulated fissionable materials will be subcritical:

$$\forall j : \frac{\sum_j A_{i,j}}{Alim_i} < 1 \tag{VI–3}$$

where
$A_{i,j}$ is the activity of isotope i in waste stream j (Bq);
and $Alim_i$ is the activity limit of disposable total activity of isotope i in waste stream j in a vault section (WER amounts) (Bq).

VI–4. CONCLUSIONS

The WAC of NRWR consist of legal requirements, international and domestic best practices, and safety case based requirements.

Three main safety goals were identified during the WAC development:

— Operational safety;
— Long term safety;
— Criticality safety.

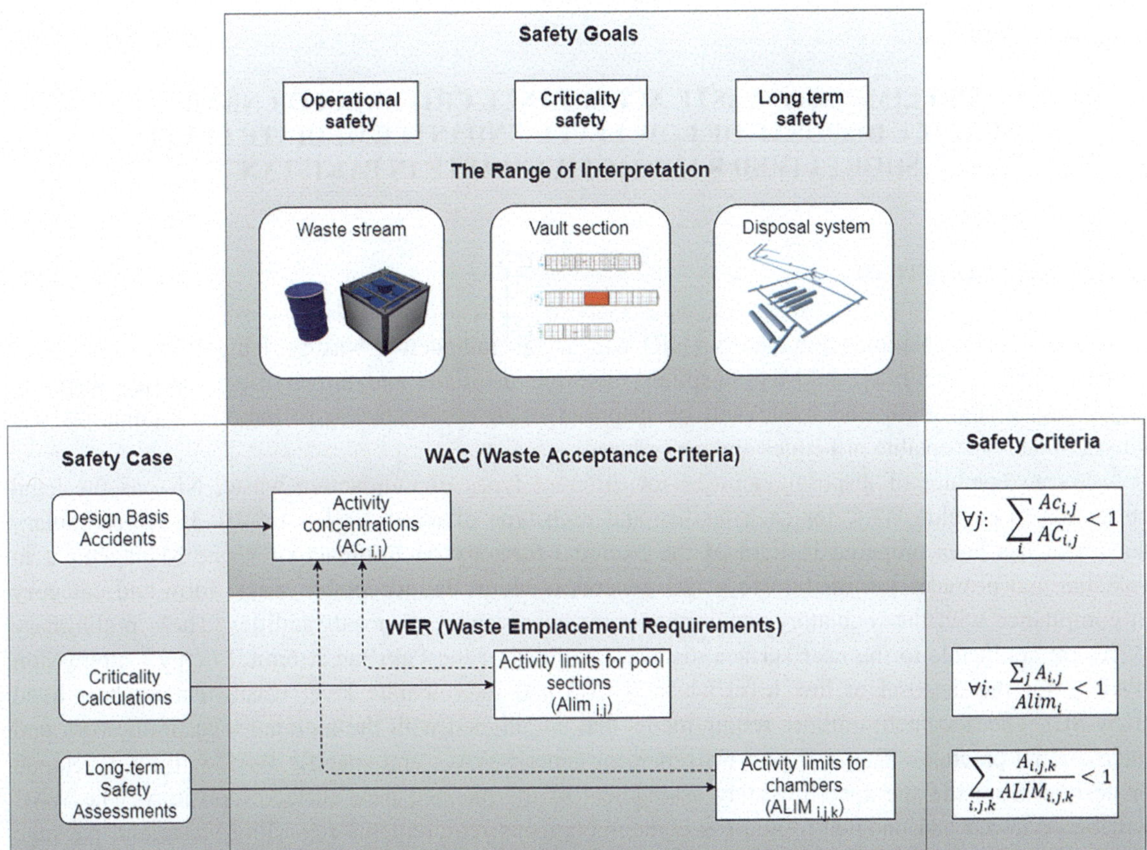

$$\forall j: \ \sum_i \frac{Ac_{i,j}}{AC_{i,j}} < 1$$

$$\forall i: \ \frac{\sum_j A_{i,j}}{Alim_i} < 1$$

$$\sum_{i,j,k} \frac{A_{i,j,k}}{ALIM_{i,j,k}} < 1$$

FIG. VI–6. Relationship between safety goals, safety analyses, waste acceptance criteria and waste emplacement requirements. Courtesy of the Public Limited Company for Radioactive Waste Management (PURAM), Hungary.

The nature of these safety goals requires different types of safety criteria and WAC (activity concentrations) and WER (activity limits) were introduced in order to define the necessary safety criteria. Safety criteria related to WAC have to be inspected during the transfer of waste packages, while safety criteria concerning WER have to be monitored during the operation of the disposal facility.

The relationship between safety goals, safety analyses, WAC, WER and the relevant safety criteria are shown in Fig. VI–6. To ensure the continuous operation of the disposal facility, the risk of accepting non-WAC compliant waste packages has to be minimized, by linking WAC and WER (dashed lines in Fig. VI–6).

REFERENCE TO ANNEX VI

[VI–1] RADIOAKTÍV HULLADÉKOKAT KEZELŐ KÖZHASZNÚ NONPROFIT, CK-TRIKOLOR, GOLDER ASSOCIATES, Bátaapáti Nemzeti Radioaktívhulladék-tároló (NRHT) Üzemeltetést megalapozó biztonsági jelentés 2020, RHK-K-005/20, Hungarian Atomic Energy Authority, Budapest (2020) (in Hungarian).

Annex VII

**PRELIMINARY WASTE ACCEPTANCE CRITERIA FOR NEAR
SURFACE DISPOSAL OF LOW LEVEL AND INTERMEDIATE LEVEL
SHORT LIVED RADIOACTIVE WASTE IN PAKISTAN**

VII–1. INTRODUCTION

The Pakistan National Repository (NR) will accept radioactive waste arising from a variety of sources (nuclear facilities, industry, hospitals, research institutes and universities) for safe disposal. Regardless of its origin, the waste will be disposed of in accordance with national regulations and internationally acceptable principles and procedures.

As an operator of disposal facilities for different types of radioactive waste, NR has the legal obligation to develop WAC for each facility and each type of waste [VII–1 to VII–3]. A preliminary document has been prepared in light of the National Strategy on Radioactive Waste Management in Pakistan and provides information to waste generators about the acceptable waste form and category in compliance with the regulatory requirements for near surface disposal facilities. These preliminary WAC are applicable to the near surface disposal facilities of the Pakistan Atomic Energy Commission (PAEC) for the disposal of low level waste (LLW) and intermediate level waste that is short lived (ILW-SL). The document outlines requirements that are aligned with the national legal framework and international practices. The requirements represent generic WAC, and specific WAC will be developed or derived from the site specific safety assessment of the planned near surface repository. The WAC will be reviewed and updated to address contemporaneous requirements regarding waste and package characteristics, certification, record keeping and the process for authorizing deviations (non-conformities) from the requirements.

VII–2. ACCEPTANCE CRITERIA FOR DISPOSAL IN NEAR SURFACE FACILITIES

Waste characteristics play an important role in the safety or safety assessment of the disposal system. Therefore, the waste packages will have to meet the specified requirements in terms of waste form, packaging, radiological and other physical and chemical characteristics. The following criteria need to be observed for the disposal of radioactive waste in near surface disposal facilities.

VII–2.1. Waste form

The criteria for the waste form are as follows:

— The class of radioactive waste will be the LLW and ILW-SL per existing Pakistan Nuclear Regulatory Authority (PNRA) Regulations [VII–1]; however, the new IAEA classification of radioactive waste [VII–4] may be used as and when adopted by PNRA.
— Only properly conditioned and solid or solidified waste will be accepted.
— The generator will provide all physical, chemical and other characteristics of the waste per WAC requirements.
— The solidified waste form will be resistant to leaching and the leach rate will comply with the standard waste form requirement.
— The waste form needs to be resistant to thermal and radiation effects.
— The compressive strength of waste form will be within relevant limits.

VII–2.2. Waste container

The criteria for the waste container are as follows:

— The waste form and container material will be compatible with each other and will be demonstrated by the generator as and when required by the operator.
— Waste will be received only in recommended (standard) containers, barrels or overpacks per designs and specifications provided to the waste generator by the operator. However, additional containers may be recommended (standardized) by stakeholders per requirements and approved by the operator.
— Waste packaged in other than recommended (standard) containers, barrels or overpacks (compatible with handling appliances) will be accepted after special approval from NR.

VII–2.3. Waste package

VII–2.3.1. Handling features

The design features of the package will provide a simple and standardized method for movement of the package during shipping and storage or disposal operations at the repository. Additionally, waste packages will have the provision to be lifted or unloaded by a fork lifter.

VII–2.3.2. Weight

The weight of an individual waste package will not exceed 5 tonnes and the total weight of an overpack (comprising number of packages) will not exceed 10 tonnes.

VII–2.3.3. Durability

The package structural durability and material thickness will be such that the container will keep its integrity during handling, interim storage and the early operational phases of disposal in a repository.

VII–2.3.4. Strength

During stacking and covering operations (several layers) waste packages need to support other waste packages without being crushed or damaged. Each waste package will be sufficiently strong to keep its integrity under normal loading and unloading conditions.

VII–2.3.5. Integrity

Each waste package needs to be in good condition with no visible cracks, holes, bulges, corrosion or other damage that could compromise integrity.

VII–2.4. Radiological criteria

External dose rates and surface contamination of the waste packages (or any overpack used during transport) will comply with PNRA regulations [VII–5].

Each package will comply with the activity limits set forth in PNRA regulations. However, these limits will be developed or derived from the site specific safety assessment of the planned near surface repository.

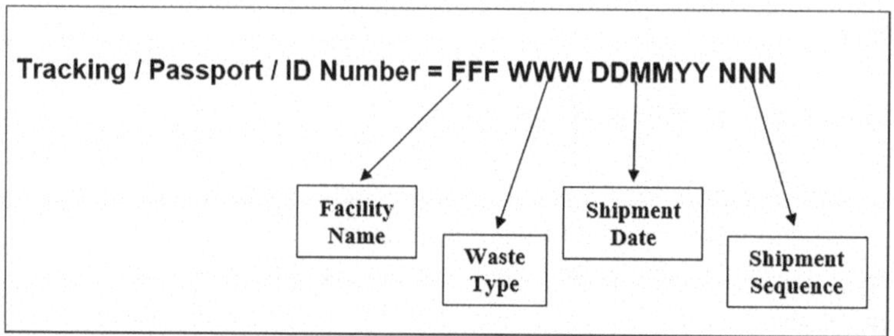

FIG. VII–1. Tracking, Passport, ID Number sequence (at the time of shipping). For example, "PIN ILS 250610 001" would indicate the first shipment from the Pakistan Institute of Nuclear Science and Technology (PINSTECH) facility of intermediate level waste that is short lived (ILW-SL) on 25 June 2010.

VII–2.5. Package identification criteria

The NR will collect waste from various generators in the country. Therefore, the establishment of a unique and self-explanatory package identification scheme is inevitable for easy traceability and management of the waste packages.

The repository operator will establish a systemic coding or identification scheme and the generator will follow this scheme accordingly (see Fig. VII–1).

Before shipment, each waste container will be properly marked and labelled, and a tracking, passport or ID number will be issued. The identification code will appear on all relevant documentation. The marking on the containers needs to be visible during inspection and at the time of shipment and until it is disposed of. Label material has to be compatible with the container material and will not impair or degrade the integrity of the package.

VII–3. WASTE THAT CANNOT BE ACCEPTED

The generator will certify that the limits given in this section are not exceeded. Certification could take place after verification by the repository operator.

VII–3.1. Waste capable of generating gases, vapours or fumes

Waste will not contain or be potentially capable of generating toxic gases, vapours or fumes harmful to workers involved in transporting, handling, disposing of and for storage and or disposal system. Waste capable of generating gases, vapours or fumes will not be accepted.

VII–3.2. Waste containing explosives or inflammable material

Waste will not be capable of detonation or reaction at normal pressures and temperatures, or of explosive composition or of explosive reaction with water or other material. Waste potentially capable of producing fire would either be excluded or made safe.

VII–3.3. Waste containing chelating agents

Waste containing more than 1% by weight of unstabilized chelating agents (amine poly-carboxylic acids, e.g. EDTA, DTPA; hydroxy-carboxylic acids and poly-carboxylic acids, e.g. citric acid, carbolic acid and gluconic acid) will not be accepted.

VII–3.4. Waste containing pyrophoric materials

Waste containing pyrophoric materials (e.g. elemental uranium, thorium and zirconium fines and turnings, powdered magnesium, calcium and titanium, phosphorous, lithium, sodium, potassium, hafnium, strontium) will not be accepted. However, waste containing pyrophoric substances may be accepted after adequate treatment, preparation and packaging as non-flammable waste.

VII–3.5. Waste containing liquid

Solid waste containing more than 1% volume of free liquid will not be accepted.

VII–3.6. Waste containing infectious material

Waste containing infectious or potentially infectious material or pathogens will not be accepted unless properly treated.

VII–3.7. Corrosive material

Waste containing any corrosive material(s) which may react with the container will not be accepted.

VII–4. NON-COMPLIANT CASES

Waste packages not complying with WAC (due to their size, form, radioactivity, etc.) can be accepted for disposal if the generator satisfactorily demonstrates that they will not violate the operational and long term safety of the disposal facility. The safety demonstration needs to be justified, documented and reported to the PNRA.

REFERENCES TO ANNEX VII

[VII–1] PAKISTAN NUCLEAR REGULATORY AUTHORITY, Regulations on Radioactive Waste Management (PAK/915), PNRA, Islamabad (2005),
http://www.pnra.org/regulations.asp#regLink

[VII–2] PAKISTAN ATOMIC ENERGY COMMISSION, National Strategy on Radioactive Waste Management (Rev.2), (Draft) PAEC, Islamabad (2007).

[VII–3] INTERNATIONAL ATOMIC ENERGY AGENCY, Disposal of Radioactive Waste, IAEA Safety Standards Series No. SSR-5, IAEA, Vienna (2011).

[VII–4] INTERNATIONAL ATOMIC ENERGY AGENCY, Classification of Radioactive Waste, IAEA Safety Standards Series No. GSG-1, IAEA, Vienna (2009).

[VII–5] PAKISTAN NUCLEAR REGULATORY AUTHORITY, Regulations for the Safe Transport of Radioactive Material (PAK/916), April 2007, Islamabad.

WASTE ACCEPTANCE CRITERIA DEVELOPMENT METHODOLOGY: THE APPROACH FOR THE CONCEPT FOR THE BOREHOLE DISPOSAL OF DISUSED SEALED RADIOACTIVE SOURCES

VIII–1. INTRODUCTION

Sealed radiation sources are widely used throughout the world and are an indispensable part of modern life that significantly contributes to the health, safety and well-being of humankind. Due to the lack of a proper nuclear infrastructure, over the years, several situations have been notified in which the control over radioactive sources has been lost. The IAEA Code of Conduct on the Safety and Security of Radioactive Sources [VIII–1] recommends the return of disused sealed radiation sources (DSRS) to the manufacturer.

There are many cases when this cannot be put in practice, therefore several African Member States requested the IAEA to respond to this situation by developing a simple and economic yet safe means of disposal for DSRS. The result is a concept where DSRS are disposed of in a narrow borehole. The first design for this conceptual disposal system was developed by the South African Nuclear Energy Corporation (NECSA) and is based on a high level of physical containment provided by placing the DSRS inside a capsule within a container (both made of stainless steel); putting these into a borehole and then backfilling with a grout based on ordinary Portland cement. Depending on the DSRS inventory and site characteristics, these stainless steel capsules and containers can contain the radionuclides for hundreds to possibly thousands of years, sufficient enough to allow the vast majority of DSRS to decay to exemption levels.

The technology is simple, and the safety assessments indicate that the system is capable of providing the required level of operational [VIII–2] and post-closure safety [VIII–3 to VIII–5]. Generic WAC have been derived for α, β, γ and neutron sources using either the lightly shielded conditioning unit or the hot cell.

VIII–2. WASTE ACCEPTANCE CRITERIA FOR THE BOREHOLE DISPOSAL CONCEPT

VIII–2.1. Components of the system

For the DSRS borehole disposal system (Fig. VIII–1), the required components are standard and are briefly described below:

— Stainless steel container with a concrete insert into which stainless steel capsule (sources are inside the capsule) fits;
— Borehole along which the containers are equally spaced (at approx. 1 m) with a plug at the bottom and a closure zone of tens or hundreds of metres deep;
— Borehole casing;
— The void between the casing and borehole wall, filled with concrete backfills;
— The void between the containers and the casing, filled with concrete backfills;
— Depending on the DSRS activity, for the source transfer into capsules and their welding, a lightly shielded conditioning unit or mobile hot cell will be used;
— A transfer flask will be used for the transfer of the containers to the borehole.

The conditioning units and the transfer flask are derived from the operational safety requirements and therefore they are providing the needed level of shielding [VIII–6]). Usually when handling the

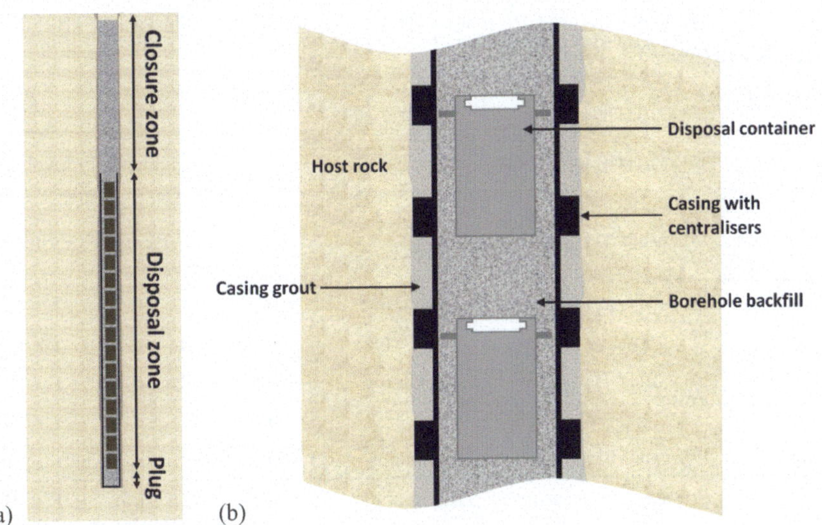

FIG. VIII–1.   (a) Layout of the disposal borehole consisting of the disposal zone and the closure zone; (b) cross-section of the disposal zone.

Category 1 and 2 sources, a hot cell is required; for some Category 3, 4 and 5 DSRS, light shielding and well developed procedures would suffice in most cases. The other items are primarily provided for post-closure safety, which relies mostly on absolute physical containment.

VIII–2.2. Identification of key parameters

VIII–2.2.1. General

When developing the WAC, general requirements refer to the complete source characterization and status (leaking or not), detailed specifications for the container and capsule, certification and proper labelling. If the source is delivered for conditioning within the machine or in a complex shielding, then is advisable that upon transfer, also the drawings to facilitate the safe dismantling and removal of the source be provided.

VIII–2.2.2. Administrative

When the source is transferred from the user to the waste processor it is generally accompanied by the original documentation which contains important information such as a complete description of the device and identification numbers. If the original documentation is not available anymore due to different reasons (lost or partially unrecoverable), then it will be necessary to recreate it using, for example, the IAEA catalogue of sealed radiation sources [VIII–7] and mandatory to characterize the source [VIII–8]. Once the paper record is created, the disposal containers are engraved with all the relevant data and specific symbols, such as date, radionuclides, activity and radiation trefoil.

VIII–2.2.3. Radionuclide inventory per container

The radionuclide content of an individual container may be limited by the heat output of the sources and the maximum doses permissible to operators and to the public post-closure. These are discussed in more detail in Section VIII–2.2.4.

The conditioning unit has to be provided with shielding that is suitable for the type of sources to be handled (e.g. lead for gamma sources, polyethylene for neutron sources). A clear strategy needs to be developed and followed related to the types of sources to be emplaced in the same capsule, their geometry and activity limits.

TABLE VIII–1. TYPICAL MAXIMUM PERMISSIBLE ACTIVITY FOR BOREHOLE DISPOSAL FOR VARIOUS ENVIRONMENTS

Based on GSA (Ref. [VIII–4])

Near field	Saturated geosphere				Total activity level (Bq)			
Disposal zone	Flow rate	Flow type	Sorption coefficients	Redox conditions	I-129	Ra-226	Pu-238	Am-241
Unsaturated	High	Porous	Low	Oxidizing	8×10^{08}	3×10^{11}	7×10^{12}	4×10^{13}
		Fractured	Low	Oxidizing	4×10^{07}	2×10^{12}	2×10^{13}	2×10^{13}
	Medium	Porous	Low	Oxidizing	8×10^{08}	7×10^{16}	1×10^{13}	5×10^{13}
			High	Reducing	4×10^{07}	*	7×10^{15}	7×10^{17}
	Low	Porous	High	Reducing	3×10^{08}	*	*	*
Saturated	High	Porous	Low	Oxidizing	3×10^{07}	2×10^{08}	1×10^{12}	2×10^{12}
		Fractured	Low	Oxidizing	3×10^{07}	5×10^{08}	2×10^{12}	2×10^{12}
	Medium	Porous	Low	Oxidizing	2×10^{07}	9×10^{13}	1×10^{13}	8×10^{12}
			High	Reducing	2×10^{07}	6×10^{17}	1×10^{18}	*
	Low	Porous	High	Reducing	6×10^{09}	*	*	*

* Indicates an activity greater than 1×10^{18} Bq.

VIII–2.2.4. *Radionuclide inventory for the repository*

The Generic Safety Assessment (GSA) [VIII–4] aims to present a post-closure safety assessment for the DSRS borehole disposal concept that is not site specific. Some results from the GSA are shown in Table VIII–1 highlighting the maximum permissible activity that could be placed in a disposal borehole for various disposal environments. The data presented Table VIII–1 are for some radionuclides with long half-lives (for radionuclides with half-lives of less than about 30 years, activities are effectively unlimited), and they highlight that containment is best achieved under saturated, low flow and chemically reducing conditions.

VIII–2.3. Quantification of acceptable limits

As mentioned in Section VIII–2.2.3, the activity limit per waste container is established based by one of the following:

— The maximum permissible dose to the operator during predisposal handling (dose limit or constraint);
— The maximum permissible dose to the public in the post-closure period (dose limit or target);
— The heat output.

VIII–2.3.1. *Dose to the operator during predisposal or disposal activities*

The external doses to the operator are due to the handling of gamma and neutron sources and, in some of the cases (if they produce bremsstrahlung radiation), beta sources. Therefore, some elements that need to be considered in the assessment are as follows:

— The strength of the DSRS and the type of radiation that it emits;
— The distance of the operator from the source and the exposure time;
— The shielding thickness and dose target.

In the case of the DSRS borehole disposal system, considerations of operator dose will lead to two sets of acceptance criteria: one for the lightly shielded facility and another for the hot cell.

In establishing the activity limits per container for a particular or a mix of radionuclide(s), it is important to establish which are the operations to be performed and the equipment that will be involved in these operations.

When estimating limits for the size of DSRS that could be handled in the lightly shielded facility, for example, it was observed that most of the operator dose occurred during the relatively short time periods (a few seconds only) when the DSRS were being transferred unshielded between the various workstations [VIII–9]. Such information is only known from trials and practical experience [VIII–10].

The dose target, in the range 2 to 10 mSv, needs to be included in the radioprotection plan. Based on the knowledge of various parameters, the operators' doses will be calculated and consequently reasonably minimized by using the hot cells for high activity sources, establishing timely limited shifts or involving a larger number of operators to implement the dose sharing principle.

VIII–2.3.2. Dose to the public in the post-closure period

The GSA activity limits for the radionuclides are derived based on the assumption that a single borehole will contain 50 disposal containers. In a generic situation, the activity per borehole is then divided by 50 to obtain the total activity limit per package. Knowledge of on-site characterization and a comprehensive up to date inventory will allow for better calculations and will avoid the over conservative approach.

VIII–2.3.3. Heat output

During the post-closure phase, the temperature of the container needs to be limited, due to the potential effects that this could have on its corrosion resistance. Such effects can be induced, for example, by the excessive evaporation of water and consequent buildup of aggressive salts. From considerations of heat transfer, this may be transformed into a limit on the decay energy (and thus activity) for the specified radionuclides (the alpha-emitting radionuclides with decay energies around 4–6 MeV are the most relevant). For the borehole disposal, it was decided that a temperature up to 50°C is acceptable and based on steady-state temperatures (applicable to long-lived radioisotopes), a maximum energy of around 50 W was established and the alpha activity was limited to 1×10^{13} Bq [VIII–4]. The calculation is very conservative for shorter-lived radionuclides, but as they are usually not alpha emitters, this makes little difference in practice.

VIII–2.4. Quality assurance

VIII–2.4.1. Development of acceptable methods for calculation, measurement and verification of parameter values

Once the operations commence, the measured doses will be checked against the calculated doses. If they are found to be less than those calculated, then the safety assessment could be re-evaluated to see if the activity limits per package can be increased.

The key parameters in the post-closure safety assessment are the site characteristics such as the geology, hydrogeology and geochemistry. The techniques involved in such studies are documented in several publications related to the deep geological disposal projects (e.g. [VIII–11, VIII–12]) but less so for the smaller scale operations that would be required for DSRS disposal in a borehole [VIII–13]. Safety assessment will include demonstrations of the robustness of the outputs through methods such as sensitivity and uncertainty analysis [VIII–14, VIII–15].

VIII–2.4.2. Development of procedures for dealing with departures or non-conformities

WAC for the DSRS borehole disposal system are significantly simpler than in most other situations. Possible non-conformities relating to WAC are:

— Incomplete inventory (discrepancies or absence of relevant information on the DSRS that need to be conditioned);
— Oversized DSRS (the geometry is not compatible with the capsule dimensions);
— Leak of the capsule, due to unproper welding or capsule defects.

Such situations are documented in the predisposal operating procedures (e.g. [VIII–16, VIII–17]).

VIII–2.4.3. Documenting the requirements and obtaining approval of the waste acceptance criteria

The required management systems standard for the DSRS borehole disposal system is similar to any other activity or facility in the field [VIII–18, VIII–19]. Therefore, the licensing documentation will need to describe the WAC development process and the mechanism that are envisaged to be implemented for checking compliance with them. Derivation of WAC will rely on (without being limited to) the safety assessment, quality assurance documents and international best practices. Compliance will be documented based on the DSRS original documents, DSRS characterization and report on the satisfactory DSRS encapsulation and containerization campaign.

VIII–3. CONCLUSIONS

The application of the DSRS borehole disposal borehole system can provide a simple and effective disposal solution for disused sealed radioactive sources. The safety is based on physical containment for all radionuclides. It can be considered a robust multibarrier system that will minimize human interaction found at inadequate storage facilities of spent sealed sources.

The DSRS borehole disposal was first proposed in 1995 for countries that do not have the possibility to co-dispose those sources with other radioactive waste. The borehole disposal concept consists of removing the sources from their devices and conditioning them into stainless steel capsules and disposal containers. These are then emplaced in one or more narrow boreholes which are backfilled and sealed using a suitable cementitious backfill material.

The design uses safe and reliable, economic and readily available materials and technology. The drilling and casing of boreholes several hundred or more metres deep has been performed worldwide for the drilling of water wells or the exploitation and exploration of natural resources. The backfilling of boreholes with cementitious materials has also been routinely done, for example, in the oil industry.

The concept requires a limited land area and infrastructure and can be implemented in a relatively short time. Over the last two decades, this disposal concept has evolved into a well developed disposal solution. Today DSRS borehole disposal projects are under way in Malaysia and are in an incipient phase in Ghana.

REFERENCES TO ANNEX VIII

[VIII–1]   INTERNATIONAL ATOMIC ENERGY AGENCY, Code of Conduct on the Safety and Security of Radioactive Sources, IAEA/CODEOC/2004, IAEA, Vienna (2004).
[VIII–2]   HANKE, J.J., Generic HAZOP for Borehole Disposal of Spent Radioactive Sources, Report No. GEA-1632, HSE Department, South African Nuclear Energy Corporation, Pretoria (2005).

[VIII–3]   KOZAK, M.W., STENHOUSE, M.J., VAN BLERK, J.J., Borehole Disposal of Spent Sources, Volume II, Preliminary Safety Assessment of the Disposal Concept, Report No. GEA-1352 (NWS-RPT-00/013), South African Nuclear Energy Corporation, Pretoria (2000).

[VIII–4]   LITTLE, R., VAN BLERK, J.J., WALK, R., BOWDEN, A. Generic Post-Closure Safety Assessment and Derivation of Activity Limits for the Borehole Disposal Concept. Report No. QRS-1128A-6 v1.0. Quintessa, Henley-on-Thames, United Kingdom (2003).

[VIII–5]   INTERNATIONAL ATOMIC ENERGY AGENCY, Generic Post-closure Safety Assessment for Disposal of Disused Sealed Radioactive Sources in Narrow Diameter Boreholes, IAEA-TECDOC-1824, IAEA, Vienna (2017).

[VIII–6]   INTERNATIONAL ATOMIC ENERGY AGENCY, Categorization of Radioactive Sources, IAEA Safety Standards Series No. RS-G-1.9, IAEA, Vienna (2005).

[VIII–7]   INTERNATIONAL ATOMIC ENERGY AGENCY, Identification of Radioactive Sources and Devices, IAEA Nuclear Security Series No. 5, IAEA, Vienna (2007).

[VIII–8]   INTERNATIONAL ATOMIC ENERGY AGENCY, Locating and Characterizing Disused Sealed Radioactive Sources in Historical Waste, IAEA Nuclear Energy Series No. NW-T-1.17, IAEA, Vienna (2008).

[VIII–9]   CROSSLAND, I.G., Waste Acceptance Criteria for the Borehole Disposal Concept, Report No. GEA 1714, HSE Department, South African Nuclear Energy Corporation, Pretoria, South Africa (2006).

[VIII–10] HANKE, J.J., Prospective hazard assessment of a borehole disposal facility, NECSA Report GEA 1633, HSE Department, South African Nuclear Energy Corporation, Pretoria, South Africa (2003).

[VIII–11] INTERNATIONAL ATOMIC ENERGY AGENCY, Experience in Selection and Characterization of Sites for Geological Disposal of Radioactive Waste, IAEA-TECDOC-991, IAEA, Vienna (1997).

[VIII–12] SVENSK KÄRNBRÄNSLEHANTERING, Site Description of Forsmark at Completion of the Site Investigation Phase - SDM-Site, SKB Report TR-08-05, Svensk Kärnbränslehantering, Stockholm (2008).

[VIII–13] INTERNATIONAL ATOMIC ENERGY AGENCY, Borehole Disposal Facilities for Radioactive Waste, IAEA Safety Standards Series No. SSG-1 (Rev. 1), IAEA, Vienna (2024),
https://doi.org/10.61092/iaea.hhr9-xnr3

[VIII–14] INTERNATIONAL ATOMIC ENERGY AGENCY, Safety Assessment Methodologies for Near Surface Disposal Facilities, IAEA, Vienna (2004).

[VIII–15] INTERNATIONAL ATOMIC ENERGY AGENCY, Safety Assessment for Facilities and Activities, IAEA Safety Standards Series No. GSR Part 4 (Rev. 1), IAEA, Vienna (2016).

[VIII–16] NEL, B. vd L., Design for the Borehole Disposal Concept, Report No. GEA-1623, HSE Department, South African Nuclear Energy Corporation, Pretoria (2003).

[VIII–17] NEL, B. vd L., Road Map and Procedures to implement the Borehole Disposal Concept, Report No. GEA-1626, HSE Department, South African Nuclear Energy Corporation, Pretoria (2004).

[VIII–18] INTERNATIONAL ATOMIC ENERGY AGENCY, The Management System for the Development of Disposal Facilities for Radioactive Waste, IAEA Nuclear Energy Series No. NW-T-1.2, IAEA, Vienna (2011).

[VIII–19] INTERNATIONAL ATOMIC ENERGY AGENCY, Leadership, Management and Culture for Safety in Radioactive Waste Management, IAEA Safety Standards Series No. GSG-16, IAEA, Vienna (2022).

GLOSSARY

The following terms used in this publication are defined in accordance with the terminology used in the IAEA Nuclear Safety and Security Glossary.[1]

departures. Departures are generally wastes or waste forms that fall outside the waste acceptance criteria and for which exceptions can be made depending upon discussions with the stakeholders (either as an exemption or a relaxation of requirements). Departures need to be identified before waste packages are produced.

generic requirements. Generic requirements are performance based requirements reflecting general conditions and the characteristics of waste packages and do not depend on the design of waste management facilities and the waste packaging process.

non-conformity. Non-conformities arise during radioactive waste management practices due to unauthorized or poorly controlled actions and/or circumstances that result in a waste package not meeting the requirements of the waste acceptance criteria.

specific requirements. Specification based requirements are derived from the detailed facility design and safety assessment and take into account specific constraints resulting for example from the nature of waste packages and the packaging process, the technical details of the facility design, the geological features of surrounding rock, etc. These requirements are closely related to the facility details and will evolve as the facility design and safety assessment progresses.

waste acceptance criteria. Quantitative or qualitative criteria specified by the regulatory body or specified by an operator and approved by the regulatory body, for the waste form and waste package to be accepted by the operator of a waste management facility.

- Waste acceptance criteria specify the radiological, mechanical, physical, chemical and biological characteristics of waste packages and unpackaged waste.
- Waste acceptance criteria might include, for example, restrictions on the activity concentration or total activity of particular radionuclides (or types of radionuclides) in the waste, on their heat output or on the properties of the waste form or of the waste package.
- Waste acceptance criteria are based on the safety case for the facility or are included in the safety case as part of the operational limits and conditions and controls.
- Waste acceptance criteria are sometimes referred to as waste acceptance requirements.

Most commonly, waste acceptance criteria are the conditions imposed on a waste producer (or an intermediary radioactive waste management organization) by the regulator and/or operator of a facility involved in the receipt of waste for handling, transportation, storage, processing and/or disposal.

Note that in this publication, the term *waste acceptance requirements* refer to specific, detailed conditions that a product, process, or facility needs to meet, and they are typically expressed as tasks, functions or constraints that have to be fulfilled, while *waste acceptance criteria* are standards, principles, or benchmarks used to evaluate or judge whether a product, process or facility meets the requirements or performs as expected.

waste form. Waste in its physical and chemical form after treatment and/or conditioning (resulting in a solid product) prior to packaging. The waste form is a component of the waste package.

waste package. The product of conditioning that includes the waste form and any container(s), in addition to any internal barriers or materials (e.g. absorbing or swelling materials and liners used for structural integrity) that might be present. A waste package is to be prepared in accordance with requirements for handling, transport, storage and/or disposal.

waste package specifications. The set of quantitative requirements to be satisfied by a waste package for handling, transport, storage and disposal.

---

[1] INTERNATIONAL ATOMIC ENERGY AGENCY, IAEA Nuclear Safety and Security Glossary, IAEA, Vienna (2022), https://doi.org/10.61092/iaea.rrxi-t56z

ABBREVIATIONS

CWP	compact waste package
DCF	dose conversion factors
DSRS	disused sealed radiation sources
DU	disposal units
GSA	Generic Safety Assessment
HLW	high level waste
HW	hazardous waste
ILW	intermediate level waste
IW	inert waste
L&C	limits and conditions
LILW	low and intermediate level waste
LLW	low level waste
NHW	non-hazardous waste
NR	national repository
NRA	nuclear regulatory authority
NRWR	National Radioactive Waste Repository
PB	producer book
QA	quality assurance
QMS	quality management system
RCA	review committee for acceptance
R&D	research and development
RWM	radioactive waste management
VLLW	very low level waste
VSLW	very short lived waste
WAC	waste acceptance criteria
WER	waste emplacement requirements

CONTRIBUTORS TO DRAFTING AND REVIEW

Antus, A.	MVM Paksi Atomeromu Zrt, Hungary
Balabanov, V.	Ignalina Nuclear Power Plant, Lithuania
Beyleveld, C.	South African Nuclear Energy Corporation, South Africa
Bogatko, A.	Ignalina Nuclear Power Plant, Lithuania
Boler, S.A.	Radioactive Waste Management Consultant, United Kingdom
Boniface, J.M.	AREVA, France
Brennecke, P.	Bundesamt für Strahlenschutz, Germany
Bruno, G.	International Atomic Energy Agency
Buday, P.	Public Limited Company for Radioactive Waste Management, Hungary
Carette, A.	Commissariat à l'énergie atomique, France
Cross, M.	Consultant, United Kingdom
Crossland, I.G.	Crossland Consulting, United Kingdom
De Dios Fisher, R.	International Atomic Energy Agency
Degnan, P.J.	Consultant, Australia
Drace, Z.	Consultant, Canada
Garamszeghy, M.	Nuclear Waste Management Organization, Canada
Herod, M.	Canadian Nuclear Safety Commission, Canada
Howell, E.K.	Facilia Projects, Austria
Johansson, C.	Swedish Nuclear Fuel and Waste Management Company, Sweden
Karakelle, B.	Turkish Atomic Energy Authority, Türkiye
Kashiwagi, M.	JGC Corporation, Japan
Krasny, D.	JAVYS — Nuclear and Decommissioning Company, Slovakia
Laraia, M.	International Atomic Energy Agency
Leganés, J.L.	Enresa, Spain
Maillard, J.L.	L'Agence nationale pour la gestion des déchets radioactifs, France
Morales Leon, A.	Consultant, Spain
Navarro, M.	Enresa, Spain
Nepeypivo, M.	Rostechnadzor, Russian Federation
Nos, B.	Public Limited Company for Radioactive Waste Management, Hungary

Ojovan, M.I.   Consultant, United Kingdom

Ormai, P.   Consultant, Hungary

Robbins, R.A.   International Atomic Energy Agency

Solente, N.   L'Agence nationale pour la gestion des déchets radioactifs, France

Tison, J.L.   L'Agence nationale pour la gestion des déchets radioactifs, France

Van Zaelen, G.   National Agency for Radioactive Waste and Enriched Fissile Material, Belgium

Voinis, S.A.   L'Agence nationale pour la gestion des déchets radioactifs, France

Welbergen, J.   Central Organisation for Radioactive Waste, Kingdom of the Netherlands

Yanigahara, S.   Japan Atomic Energy Research Institute, Japan

Zavazanova, A.   Nuclear Regulatory Authority of the Slovak Republic, Slovakia

Consultants Meetings

Vilnius, Lithuania: 9–13 November 2009

Vienna, Austria: 10–14 May 2010

Technical Meeting

Vienna, Austria: 23–27 November 2020

Structure of the IAEA Nuclear Energy Series*

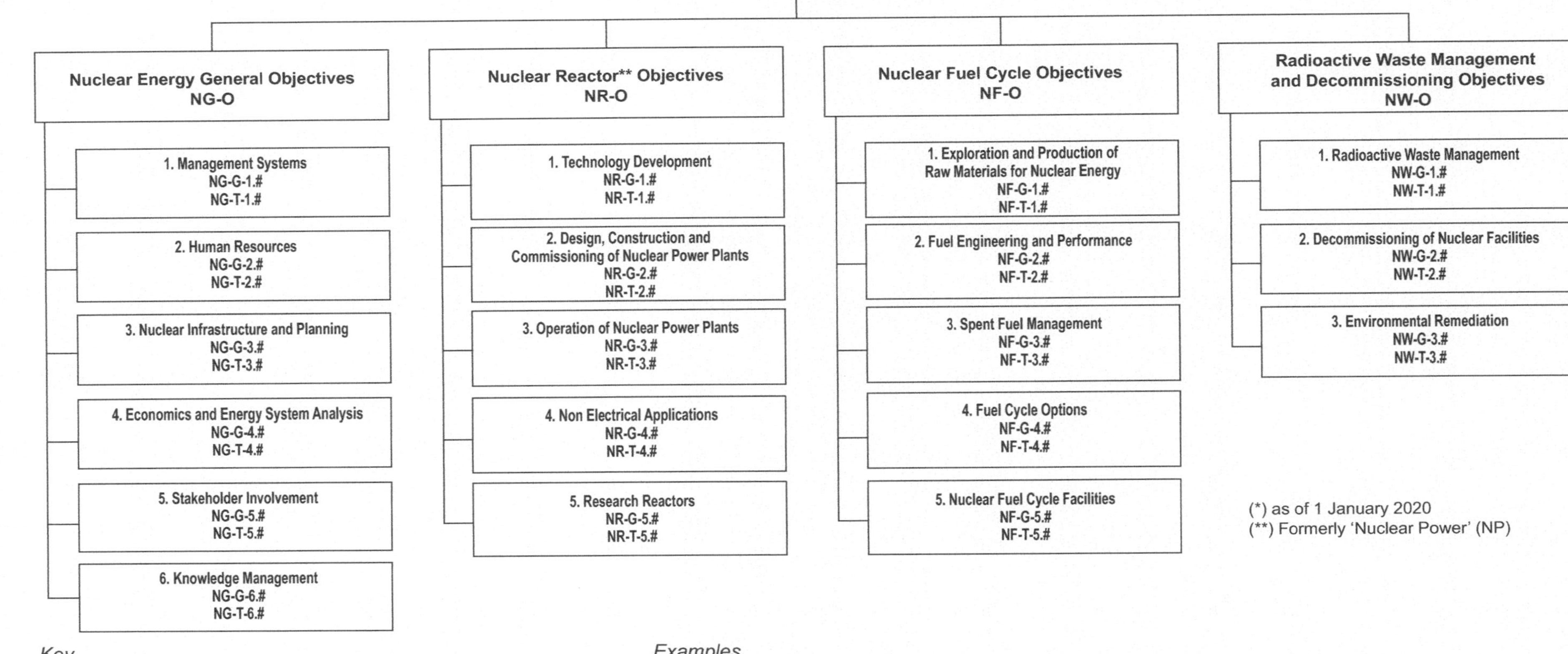

CONTACT IAEA PUBLISHING

Feedback on IAEA publications may be given via the on-line form available at:
www.iaea.org/publications/feedback

This form may also be used to report safety issues or environmental queries concerning IAEA publications.

Alternatively, contact IAEA Publishing:

Publishing Section
International Atomic Energy Agency
Vienna International Centre, PO Box 100, 1400 Vienna, Austria
Telephone: +43 1 2600 22529 or 22530
Email: sales.publications@iaea.org
www.iaea.org/publications

Priced and unpriced IAEA publications may be ordered directly from the IAEA.

ORDERING LOCALLY

Priced IAEA publications may be purchased from regional distributors and from major local booksellers.

Printed and bound by CPI Group (UK) Ltd, Croydon, CR0 4YY

06/07/2026

02160600-0011